Introduction to the Mathematical Physics of Nonlinear Waves (Second Edition)

Introduction to the Mathematical Physics of Nonlinear Waves (Second Edition)

Minoru Fujimoto
University of Guelph, Ontario, Canada

IOP Publishing, Bristol, UK

ISBN 978-0-7503-3759-5 (ebook)
ISBN 978-0-7503-3757-1 (print)
ISBN 978-0-7503-3760-1 (myPrint)
ISBN 978-0-7503-3758-8 (mobi)

DOI 10.1088/978-0-7503-3759-5

Version: 20211001

IOP ebooks

British Library Cataloguing-in-Publication Data: A catalogue record for this book is available from the British Library.

Published by IOP Publishing, wholly owned by The Institute of Physics, London

IOP Publishing, Temple Circus, Temple Way, Bristol, BS1 6HG, UK

US Office: IOP Publishing, Inc., 190 North Independence Mall West, Suite 601, Philadelphia, PA 19106, USA

Minoru Fujimoto, 1490 Rathburn Rd E *301, Mississauga ON L4W 4Z8, Canada

To the memory of Toshiya Taniuchi

Contents

Notes on the second edition

Nonlinear dynamics is a well-established discipline in physics today, and this edition offers a new comprehensive account of the basic soliton theory revised for crystalline processes. The content has been changed from the previous edition to emphasize thermodynamics to provide a proper description of the modulated crystals. Accordingly, nonlinear phenomena in crystals are discussed for students to learn the subject matter with thermodynamics in a more pleasing manner than existing references; providing stimulating material to avoid confusion. Moreover, recently discovered facts on superconductivity are discussed in this edition, providing a new view on condensed matter.

Basic mathematics of elliptic functions is a prerequisite to nonlinear physics. The necessary theory is reviewed in chapter 1 for those who are not particularly familiar with it, while Wikipedia provides useful references of elliptic and hyperbolic functions for self-study.

25 October 2020
Minoru Fujimoto

Preface to the first edition

Nonlinear problems can be analyzed using inhomogeneous differential equations to obtain information on the nonlinear content, which was harder to do using traditional approaches. Nevertheless, recent mathematical studies have revealed that equations of Sturm–Liouville's type can be specified by soliton solutions, representing potential energies of the surrounding medium, subject to the law of conservation of energy.

Nonlinearity arises from the dynamical response of surroundings, which is essentially related to boundaries and interactions. Properties of a medium can therefore be analyzed from mathematical consequences; however, the thermodynamic environment must also be considered, whose unspecifiable conditions make it difficult to interpret. Nonlinear equations in crystals are generally restricted by thermodynamic conditions that are idealized with external pressure and temperature. Justified by the *least action principle*, the Hamilton equations represent for conservative dynamical systems to be evaluated with the minimized Gibbs' potential. In practice, singularities arising from nonlinear dynamics, associated with entropy production in crystalline processes to the following equilibrium states.

In equilibrium, the dynamical system is specified by eigenvalues of the Hamiltonian function, called a *canonical ensemble*, and thermodynamic processes are known as *canonical transformations* in statistical mechanics. In this viewpoint, the soliton theory substitutes canonical uncertainty, similar for the rigid lattice to be substituted by vibrational phonon stability, and Born–Huang's transitions cover adiabatic fluctuations. In this book, the soliton dynamics appear for canonical transitions to be discussed by soliton scatterings.

This book was written for students at advanced undergraduate and graduate levels to be used as a textbook on nonlinear physics. Elliptic and hyperbolic functions are prerequisites to such courses, while most students are not familiar with that mathematics. I have therefore included a short discussion on these functions in chapter 1 for readers' convenience.

Acknowledgements

As an experimental scientist myself, I have enormously benefitted from *Elements of Soliton Theory* and *Lecture Notes on Nonlinear Waves* in writing this book. My sincere thanks are therefore expressed to Professor G L Lamb Jr and Professors T Taniuchi and K Nishihara. Also, I thank my wife Haruko for her continuous encouragement.

August 2013
Minoru Fujimoto

Author biography

Minoru Fujimoto

Minoru Fujimoto is a retired professor from the University of Guelph, Ontario, Canada. During his association with the university, his research area was in the field of magnetic resonance studies on structural phase transitions in crystals, which has currently been extended to theoretical work with soliton dynamics. He is the author of several books, including *Physics of Classical Electromagnetism* and *Thermodynamics of Crystalline States* (Springer); *Introduction to Mathematical Physics of Nonlinear Waves* and *Solitons in Crystalline Processes* (IOP Publishing). He lives in Mississauga, Ontario: mfujimotp@outlook.com.

IOP Publishing

Introduction to the Mathematical Physics of Nonlinear Waves
(Second Edition)

Minoru Fujimoto

Chapter 1

Nonlinearity and elliptic functions in classical mechanics

In this chapter, we discuss the origin of nonlinearity that is often dismissed in classical mechanics as insignificant or bypassed by using idealized models to simplify mathematical analysis. Energy damping during entropy production is attributed to air friction, where the dynamical stability is evaluated by thermal relaxation. Such energy loss in microscopic physics should also be considered with respect to surrounding media for normal terrestrial phenomena. Referring to the *least action principle*, dynamical systems may not necessarily be *canonical* in general, so the compatibility of models with the environment should be determined in many cases of nonlinearity, as will be discussed in the following chapters. In classical mechanics, most environmental conditions are usually ignored for mathematical simplicity, which however need to be reconsidered for modern theories of nonlinear dynamics. Some significant examples are therefore discussed in this chapter as preliminaries for this book.

Mathematically, elliptic integrals and Jacobi's elliptic functions are heavily utilized for nonlinear dynamics, thus these are discussed in this chapter as essential mathematic preliminaries.

1.1 A pendulum

1.1.1 Oscillations

An oscillating mass m of a pendulum in a gravitational field is a standard problem in classical mechanics, providing the principle for mechanical stability. Attributed to work by restoring gravitational force, its excitation at a small amplitude is sufficient to evaluate stability at a minimum potential energy. At a finite amplitude however,

doi:10.1088/978-0-7503-3759-5ch1

the motion cannot be steady against air friction, and ceases eventually at zero amplitude.

Figure 1.1 illustrates a simple pendulum of mass m that is hung with an inflexible string or rod of length l. Assuming the supporting point P fixed at the rigid ceiling, we can write the equation of motion with respect to coordinate axes fixed in space as

$$ml\frac{\mathrm{d}^2\phi}{\mathrm{d}t^2} = -mg\cos\phi \quad \text{and} \quad \mathrm{T} = -mg\sin\phi, \tag{1.1}$$

where $\phi(t)$ is the angle of rotation around P in the vertical plane; g and T are the gravitational accelerations and tension in the string, respectively. Further, we postulate that the motion is restricted in the vertical plane. The Earth's rotation is ignored, which nevertheless is an acceptable assumption for a normal observation within a short period of time.

Equation (1.1) is nonlinear, however expanding $\cos\phi = 1 + \frac{1}{2}\phi^2 + \cdots$ for a small ϕ, (1.1) can be reduced to a linear equation. Integrating (1.1) for $\cos\phi \sim 1$, we obtain

$$\frac{1}{2}ml^2\dot{\phi}^2 + mgl(1 - \cos\phi) = E, \tag{1.2}$$

where $E = mgl(1 - \cos\alpha)$ and $\alpha = \phi_{t=0}$. Solving (1.2), we obtain

$$\frac{\mathrm{d}\phi}{\mathrm{d}t} = 2\sqrt{\frac{g}{l}}\sqrt{\kappa^2 - \sin^2\frac{\phi}{2}} \quad \text{where } \kappa = \sin\frac{\alpha}{2}.$$

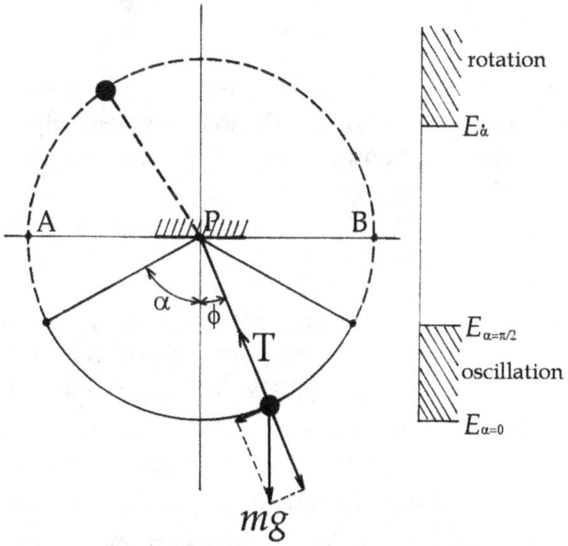

Figure 1.1. A simple pendulum.

Setting $\sin\frac{\phi}{2} = \kappa \sin\varphi$ for another angle φ, we can write

$$\mathrm{d}\phi = \frac{2\kappa\cos\varphi\,\mathrm{d}\varphi}{\sqrt{1-\kappa^2\sin^2\varphi}}.$$

Defining the time interval as the period $\mathbf{T} = \int_{\varphi=0}^{\varphi=\pi/2}\mathrm{d}t$, we have

$$\int_0^t \mathrm{d}t = \sqrt{\frac{l}{g}}\int_0^{\varphi}\frac{\mathrm{d}\varphi}{\sqrt{1-\kappa^2\sin^2\varphi}} = \sqrt{\frac{l}{g}}F(\kappa,\varphi), \quad \text{where}$$
$$F(\kappa,\varphi) = \int_0^{\varphi}\frac{\mathrm{d}\varphi}{\sqrt{1-\kappa^2\sin^2\varphi}} \tag{1.3}$$

is an elliptic integral of the first kind, indicating the repetition time specified by $0 \leqslant \varphi \leqslant 2\pi$.

The period of oscillation can therefore be defined by

$$\mathbf{T} = 4\int_0^{\pi/2}\mathrm{d}t = \sqrt{\frac{l}{g}}F\left(\kappa,\frac{\pi}{2}\right), \quad \text{where } F\left(\kappa,\frac{\pi}{2}\right) = K(\kappa) = \int_0^{\pi/2}\frac{\mathrm{d}\varphi}{\sqrt{1-\kappa^2\sin^2\varphi}};$$

$$\text{i.e. } \mathbf{T} = 4K(\kappa)\sqrt{\frac{l}{g}}. \tag{1.4}$$

Writing $z = \sin\varphi$, the inverse function $u = F(\kappa,\varphi)$ can be defined from

$$u = \int_0^{\varphi}\frac{\mathrm{d}\varphi}{\sqrt{1-\kappa^2\sin^2\varphi}} = \mathrm{sn}^{-1}z, \tag{1.5}$$

where $z = \mathrm{sn}\,u$ is *Jacobi's elliptic* sn-*function*, and the angle $\varphi = \mathrm{am}\,u$ was originally defined as the *amplitude function* of u. We note that the elliptic function $\mathrm{sn}\,u$ is periodic as $\sin\varphi$. Figure 1.2 shows a graph of u-φ, plotted against representative *modulus* κ.

1.1.2 Vertical rotation

Equation (1.2) expresses the law of energy conservation, where the first and second terms are the kinetic and potential energies, respectively. The kinetic energy $K = \frac{1}{2}ml^2\dot{\phi}^2$ is maximum at $\phi = 0$, and $K = 0$ is minimum at $\phi = \alpha$, so that the motion is restricted in the range $-\alpha \leqslant \phi \leqslant \alpha$ for a small K.

On the other hand, as shown in figure 1.1, a circular motion can take place in a vertical plane, if $K \geqslant \frac{1}{2}ml^2\dot{\alpha}^2$ is sufficiently large. Oscillatory motion can take place in this case, in the range $0 \leqslant \alpha < \frac{\pi}{2}$; no circular motion occurs, unless $\dot{\alpha} \geqslant \dot{\alpha}_{\pi}$, assuming the string to be taut. Normally, these are specified in the *initial conditions.*

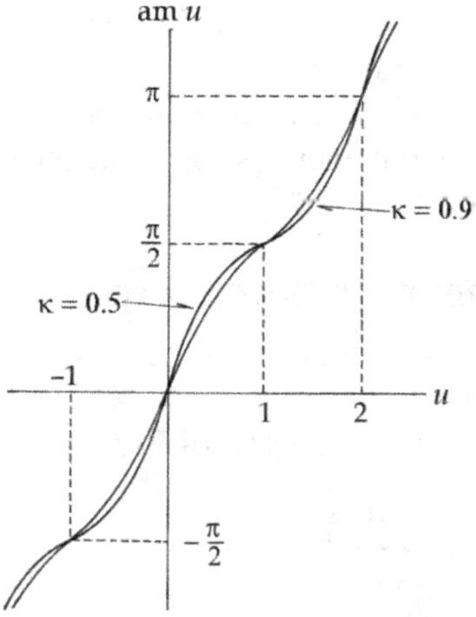

Figure 1.2. Jacobi's amplitude function $\varphi = \text{am}(u)$ for sn u.

For circular motion, the centrifugal force should be sufficiently strong, although this can be ignored for a pendulum with a small amplitude, and the centrifugal force is considered as

$$ml\ddot{\phi} = \text{T} - mg\cos\phi \approx \text{T}. \tag{1.6}$$

In the radial direction, where T is a function of $\ddot{\phi}$ that is not constant in general. We can assume $\frac{1}{2}ml^2\dot{\phi}^2 - \frac{1}{2}ml^2\dot{\alpha}^2 = \int_0^\phi \text{T}l\,\mathrm{d}\phi \geqslant 0$ for the rotating pendulum at $E \geqslant E_{\dot{\alpha}}$, where

$$E = \frac{1}{2}ml^2\dot{\phi}^2 + \text{T}\,l\phi \qquad \text{and} \qquad E_{\dot{\alpha}} = \frac{1}{2}ml^2\dot{\alpha}^2 + \text{T}l\alpha \approx 2mgl.$$

The energy criteria for oscillation and rotation are illustrated in figure 1.1, where $E > 2mgl$ is for rotation. For oscillation, we have $1 - \cos\alpha = \frac{E}{mgl}$ and $\kappa = \sqrt{\frac{E}{2mgl}}$. On the other hand, with $E > 2mgl$, the time of rotation can be defined by an elliptic integral at a different modulus

$$\kappa_0 = \sqrt{\frac{2mgl}{E}} < 1. \tag{1.7}$$

Therefore, writing

$$t = \frac{\kappa_0}{2}\sqrt{\frac{l}{g}}\int_0^\theta \frac{\mathrm{d}\theta}{\sqrt{1 - \kappa_0^2\sin^2\frac{\theta}{2}}} = \kappa_0\sqrt{\frac{l}{g}}F\left(\kappa_0, \frac{\theta}{2}\right), \tag{1.8}$$

the period of rotation is given by

$$T = 2\kappa_0\sqrt{\frac{l}{g}}K(\kappa_0). \tag{1.9}$$

In the range $2mgl > E > -mgl$, the motion cannot be fully described by this model, unless an additional condition is imposed for the string to be straight during the motion; accordingly, the system belongs to the non-canonical category. With well-defined kinetic and potential energies, the system can be classified as canonical, but remains non-canonical otherwise.

1.2 Vibration by a nonlinear spring force

Anharmonic vibration is usually discussed with a force characterized by a potential energy $V(z)$ of non-Hooke type, which can however be harmonic, if $V(z) \propto -z^2$. The nonlinearity arises from anharmonic parts of $V(z)$, as described by

$$\frac{1}{2}\left(\frac{\mathrm{d}z}{\mathrm{d}t}\right)^2 + V(z) = E \tag{1.10}$$

in one-dimension, considering the mass $m = 1$ for simplicity. Integrating (1.10), we obtain the expression

$$t = \pm\int_0^z \frac{\mathrm{d}z}{\sqrt{2\{E - V(z)\}}}, \tag{1.11}$$

where the *symmetric restoring potential energy* can be expressed as

$$V(z) = \frac{\alpha}{2}z^2 + \frac{\beta}{4}z^4. \tag{1.12}$$

As in a pendulum problem, we can find singularities of the potential (1.12) that is determined from $E = V(z)$, namely

$$E - V(z) = \frac{\beta}{2}(a - z)(a + z)(b^2 + z^2) = 0,$$

where $a^2 = \frac{-\alpha + \sqrt{\alpha^2 + 4\beta E}}{\beta}$ and $b^2 = a^2 + \frac{2\alpha}{\beta}$.

Defining a function $v(z) = \frac{\beta}{2}(b^2 + z^2)$, we write an equation

$$2\{E - V(z)\} = (a - z)(a + z)v(z)$$

in (1.11), confirming oscillations in the range $-a < z < +a$.

Writing $z = a\cos\varphi$, we have $v(z) = \frac{\beta}{2}(b^2 + a^2\cos^2\varphi)$. Therefore, the time for oscillation between $-\varphi$ and $+\varphi$ determined by (1.8) can be expressed by an elliptic integral

$$t = \sqrt{\frac{2}{\beta(a^2 + b^2)}}\int_0^\varphi \frac{\mathrm{d}\varphi}{\sqrt{1 - \kappa^2\sin^2\varphi}} \quad \text{where } \kappa^2 = \frac{a^2}{a^2 + b^2}. \tag{1.13}$$

On the other hand, if the potential $V(z)$ is *asymmetric*, the equation $E - V(z) = 0$ for $V(z) = \frac{1}{2}\alpha z^2 + \frac{1}{3}\gamma z^3$ is an algebraic equation of the 3rd order that is characterized by three roots in general in figure 1.3. Curve A in this figure is

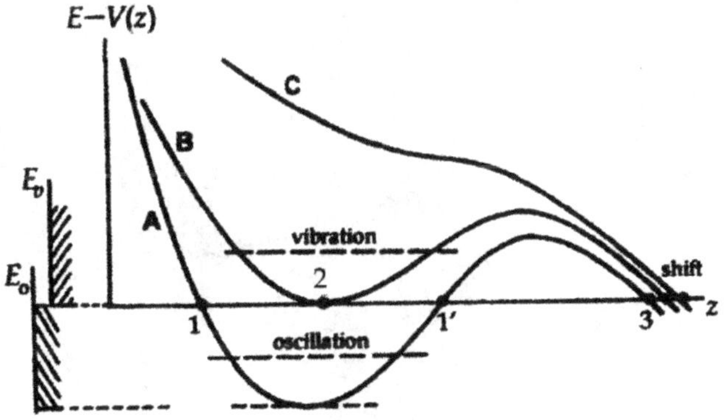

Figure 1.3. $E - V(z)$ vs. z. Roots of $E - V(z) = 0$ are indicated by 1, 1′, 2 and 3 for specific solutions.

specified by three real roots, while in curve B by two roots, both cases showing the presence of oscillatory motion. In curve C, however, there is no oscillation.

In terms of the total energy E, the oscillatory modes are separated discontinuously from non-oscillatory cases, as for a simple pendulum. The nature of the energy gap is unknown however, attributable to *transitions*, if identified as a canonical process.

1.3 Hyperbolic and elliptic functions

In the previous discussions, we used elliptic integrals to express time for nonlinear dynamics. However, the mathematics of elliptic functions may be unfamiliar to those with no exposure to nonlinear physics, for whom the basic formulas are summarized in this section. These days, excellent corrections of elliptic and hyperbolic formulas are conveniently available online from Wikipedia [1], without the need to consult older references [2].

1.3.1 Definitions

The elliptic function $z = \text{sn}\, u$ was defined as the reverse function of elliptic integral (1.5) in section 1.1.1. Here, mathematical properties of elliptic functions, known as Jacobi's functions, and related hyperbolic functions are summarized.

Understandably, the reason for defining $z = \text{sn}\, u$ is due to a similar relation between (1.5) and the integral formula for trigonometric

$$\sin^{-1} z = u = \int_0^z \frac{\mathrm{d}z}{\sqrt{1 - z^2}} \quad \text{and} \quad z = \sin u. \tag{1.14a}$$

In fact, we defined $z = \text{sn}\, u$ from the integral

$$u = \int_0^z \frac{\mathrm{d}z}{\sqrt{(1 - z^2)(1 - \kappa^2 z^2)}}. \tag{1.14b}$$

We note that (1.5) is identical to (1.14*b*) if $\kappa = 0$, whereas (1.5) at $\kappa = 1$ is expressed by a *hyperbolic function,* as indicated later. In figure 1.4 the graphic comparison of

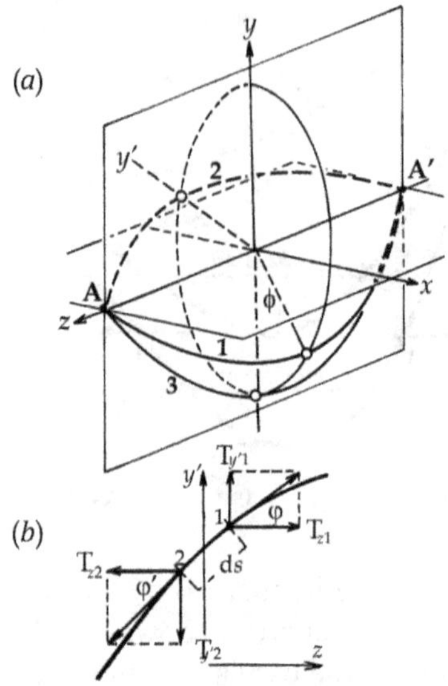

Figure 1.4. (*a*) A jumping rope. (*b*) Balanced tensions on a differential rope d*s*.

$\sin^{-1} z$ and $\text{sn}^{-1} z$ is shown. Accordingly, reverse functions of (1.14*b*) at a modulus $0 < \kappa < 1$ are expressed as $u = \text{sn}^{-1}(z, \kappa)$ and $z = \text{sn}(u, \kappa)$, which are called Jacobi's sn-functions.

Analogous to a trigonometric cos-function, we can define elliptic cn u by the relation

$$\text{sn}^2 u + \text{cn}^2 u = 1, \tag{1.15}$$

and further define

$$\text{dn}^2 u = \frac{1}{1+\kappa^2}. \tag{1.16}$$

At $\kappa = 0$ in particular, these Jacobi's functions become trigonometric, namely

$$\text{sn}(u, 0) = \sin u, \quad \text{cn}(u, 0) = \cos u \quad \text{and} \quad \text{dn}(u, 0) = 1. \tag{1.17}$$

For $0 < \kappa < 1$ elliptic functions are in a deformed shape as illustrated in figure 1.4, but for $\kappa = 1$ in contrast

$$u = \int_0^z \frac{\text{d}z}{1-z^2} = \tanh^{-1} z \quad \text{or} \quad z = \tanh u,$$

and

$$\text{sn}(u, 1) = \tanh u \quad \text{and} \quad \text{cn}(u, 1) = \text{dn}(u, 1) = \text{sech}\, u. \tag{1.18}$$

1.3.2 Differentiation

From the definition of (1.14*b*), we can write a derivative

$$\frac{du}{dz} = \frac{1}{\sqrt{(1 - z^2)(1 - \kappa^2 z^2)}}.$$

Using the relation $z = \mathrm{sn}^{-1} u$ of (1.5) to replace z in the above expression, we have

$$\frac{dz}{du} = \frac{d(\mathrm{sn}\, u)}{du} = \sqrt{(1 - \mathrm{sn}^2 u)(1 - \kappa^2 \mathrm{sn}^2 u)} = \mathrm{cn}\, u \ \mathrm{dn}\, u. \tag{1.19}$$

Similarly, we can write

$$\frac{d(\mathrm{cn}\, u)}{du} = \frac{d}{du}\sqrt{1 - \mathrm{sn}^2 u} = \frac{-\mathrm{sn}\, u}{\sqrt{1 - \mathrm{sn}^2 u}} \frac{d\, \mathrm{sn}\, u}{du} = -\mathrm{sn}\, u \ \mathrm{dn}\, u \tag{1.20}$$

$$\frac{d(\mathrm{dn}\, u)}{du} = \frac{d}{du}\sqrt{1 - \kappa^2 \mathrm{sn}^2 u} = \frac{-\kappa^2 \mathrm{sn}^2 u}{\sqrt{1 - \kappa^2 \mathrm{sn}^2 u}} \frac{d\, \mathrm{sn}\, u}{du} = -\kappa^2 \mathrm{sn}\, u \ \mathrm{cn}\, u. \tag{1.21}$$

1.3.3 Reverse functions cn^{-1} and dn^{-1}

The reverse function of sn-function is expressed in (1.5) by sn^{-1}, which is essentially an elliptic integral. Similarly, we can write such integrals from cn and dn functions.

Letting $z = \mathrm{cn}\, u$, defining $\kappa' = \sqrt{1 - \kappa^2}$ for (1.19), we can write

$$\frac{dz}{du} = \sqrt{(1 - z^2)}\sqrt{1 - \kappa^2(1 - z^2)} = \sqrt{(1 - z^2)(\kappa'^2 + \kappa^2 z^2)},$$

so that

$$u = \mathrm{cn}^{-1}(z, \kappa) = -\int_1^z \frac{dz}{\sqrt{(1 - z^2)(\kappa'^2 + \kappa^2 z^2)}}, \tag{1.22}$$

where the lower limit of the integral is $z = 1$, corresponding to $\mathrm{cn}\, u = 1$.

Similarly, we can derive

$$\mathrm{dn}^{-1}(z, \kappa') = -\int_1^z \frac{dz}{\sqrt{(1 - z^2)(z^2 - \kappa'^2)}}. \tag{1.23}$$

1.4 A jumping rope

1.4.1 The shape

We discuss here another classical example of a jumping rope, swinging or rotating around the fixed axis AA', as showing in figure 1.4(*a*). To turn the rope, an additional energy is required as given by the energy difference between these modes. It is noted that the rope in steady rotation is like the hanging rope, but in a different shape with respect to the coordinate system on the ground.

Figure 1.4(*a*) shows such a rope, swinging, rotating and hanging indicated by lines 1, 2 and 3. As remarked, these are in similar shapes but not quite identical, depending on angular frequencies ω and the mass density ρ of the rope.

To set the rope in steady rotation, a sufficiently high initial angular speed $\dot{\alpha}$ is required, while for swinging the initial angle ϕ can be sufficiently small. Gravity can be ignored for the rotation.

The rope is assumed to be flexible in a fixed length l, for simplicity. However, the tension T is not uniformly distributed on the rope, where the centrifugal force is balanced with the deformed shape. Figure 1.4(*b*) illustrates the equilibrium situation.

We define the curvilinear coordinate s along the curved rope, thereby assuming the tension as given by a function T(s), where s is a measure from one end, A, and $0 \leqslant s \leqslant l$. The rope is assumed in steady shape in the rotating frame of reference $xy'z$, the mass of the segment ρ ds at an arbitrary position should be balanced by the relations

$$\mathrm{T}_{z1}\,\mathrm{d}s = \mathrm{T}_{z2}\,\mathrm{d}s \quad \text{and} \quad \mathrm{T}_{y'1}\,\mathrm{d}s - \left\{\mathrm{T}_{y'2} - \left(\frac{\mathrm{dT}_{y'2}}{\mathrm{d}s}\right)\mathrm{d}s\right\} = \rho\omega^2 y'\,\mathrm{d}s.$$

Here, assuming $\frac{\mathrm{dT}_{y'}}{\mathrm{d}s} = \mathrm{T_o}\cos\varphi$, where $\mathrm{T_o}$ is constant, we have $\cos\varphi = \frac{\mathrm{d}y'}{\mathrm{d}z}$ for the angle φ to deform the rope from the z-axis. Therefore, we derive

$$\frac{\mathrm{d}^2 y'}{\mathrm{d}z^2} + \frac{\rho\omega^2}{\mathrm{T_o}}y' = 0, \tag{1.24}$$

allowing to consider y' to be a function of z, i.e. $y' = f(z)$.

Equation (1.24) can be solved with an elliptic integral, as explained in the following. Setting $p = f'(z)$, from the formula $\mathrm{d}s^2 = \mathrm{d}y'^2 + \mathrm{d}z^2$ we obtain $\left(\frac{\mathrm{d}s}{\mathrm{d}z}\right)^2 = 1 + p^2$. Accordingly, (1.24) can be modified as $p\frac{\mathrm{d}p}{\mathrm{d}y'} = 1 + \frac{\rho\omega^2 y'}{\mathrm{T_o}}\sqrt{1+p^2}$. Integrating this, we derive $\sqrt{1+p^2} = 1 + \frac{\rho\omega^2}{2\mathrm{T_o}}(b^2 - y'^2)$, where b is a constant determined by $p = 0$. Solving (1.24) for the constant κ, we derive $\kappa^2 = \frac{\rho\omega^2 b^2/4\mathrm{T_o}}{1+\rho\omega^2 b^2/4\mathrm{T_o}}$ for $0 < \kappa < 1$.

Then, using a new variable $\xi = y'/b$ for convenience, from the relations

$$\frac{\mathrm{d}z}{c} = \frac{\mathrm{d}\xi}{\sqrt{(1-\xi^2)(1-\kappa^2\xi^2)}}, \quad \text{where } \frac{1}{c} = \sqrt{\frac{\rho\omega^2}{\mathrm{T_o}}\left(1 + \frac{\rho^2\omega^2 b^2}{4\mathrm{T_o}}\right)} = \frac{2\kappa}{(1-\kappa^2)b},$$

we can express the reverse function

$$y' = b\,\mathrm{sn}\frac{z}{c} \quad \text{where } \frac{b}{c} = \frac{2\kappa}{1-\kappa^2}. \tag{1.25}$$

The rope's shape is specified by the terminal conditions $y' = 0$ at $z = 0$ and $z = 2a$ together with $y' = b$ at $z = a$, and expressed as a sn-function with the modulus κ.

For a hanging rope, replacing the centrifugal potential $\rho\omega^2 y'$ by the gravitational one $\rho g y\ \mathrm{d}z$, we have the expression

$$1 + \frac{\mathrm{d}}{\mathrm{d}s}\left(\mathrm{T_o}\frac{\mathrm{d}y}{\mathrm{d}s}\right) + \rho g = 0,$$

by which the rope's shape can be described as

$$y = a \cosh \frac{z - z_0}{a} - \rho g a. \tag{1.26}$$

Here, z_0 and$a$ are the center position of AA' determined by $\frac{\mathrm{d}y}{\mathrm{d}z} = 0$ and the length of the rope, respectively.

1.4.2 Periodicity of Jacobi's sn-function

Equation (1.26) describes a general form of a jumping rope with both ends fixed. Here, we discuss a rope in curve of a fixed length between $z = +a$ and $-a$, referring to figure 1.4. The sn-function is periodic along the z-axis for the distance AA', which is convenient for simplifying discussion. In any case, we differentiate (1.25) to obtain

$$\frac{\mathrm{d}y'}{\mathrm{d}z} = \frac{b}{c}\mathrm{cn}\left(\frac{z}{c}\right)\mathrm{dn}\left(\frac{z}{c}\right).$$

Accordingly,

$$\left(\frac{\mathrm{d}s}{\mathrm{d}z}\right)^2 = 1 + \frac{b^2}{c^2}\mathrm{cn}^2\frac{z}{c}\mathrm{dn}^2\frac{z}{c} = 1 + \frac{b^2}{c^2}\frac{1-\kappa^2}{\kappa^2}\mathrm{dn}^2\frac{z}{c} + \frac{b^2}{c^2\kappa^2}\mathrm{dn}^4\frac{z}{c}.$$

From (1.25) the relation $\frac{b}{c} = \frac{2\kappa}{1-\kappa^2}$, which simplifies the above to

$$\left(\frac{\mathrm{d}s}{\mathrm{d}z}\right)^2 = \left(1 - \frac{2}{1-\kappa^2}\mathrm{dn}^2\frac{z}{c}\right)^2.$$

Hence, we have

$$s(z) = \frac{2}{1-\kappa^2}\int_0^z \mathrm{dn}^2\left(\frac{z}{c}\right)\mathrm{d}z - z, \tag{1.27a}$$

that expresses the phase angle ϕ defined by $z = cu$ with $\sin\phi = \mathrm{sn}\,u$ between A and A'. It is noted that this ϕ is not the same φ in (1.13), while both represent sinusoidal phases of the function sn u, hence $\cos\phi\ \mathrm{d}\phi = (\mathrm{cn}\,u\ \ \mathrm{dn}\,u)du$.

Therefore, comparing with the relation $\cos^2\phi = 1 - \sin^2\phi = 1 - \mathrm{sn}^2 u = \mathrm{cn}^2 u$, we can write $\mathrm{d}\phi = (\mathrm{dn}\,u)\mathrm{d}u$, allowing for the following integral to deal with (1.27a). The integral

$$\int_0^u (\mathrm{dn}^2 u)\mathrm{d}u = \int_0^\phi \sqrt{1 - \kappa^2 \sin^2 \phi}\, \mathrm{d}\phi = E(\kappa, \phi) \tag{1.27b}$$

is called the *elliptic integral of the second kind*, which indicates the half-period of elliptic sn-function. Equation (1.27*b*) can then be modified as

$$s(z) = \frac{2c}{1 - \kappa^2} E\left(\frac{z}{c}\right) - z. \tag{1.27c}$$

The terminal points of the rope A and A′ signified at $z = -a$ and $+\, a$, respectively, are separated by $2K(\kappa)$ that is defined by

$$K(\kappa) = \int_0^\pi \sqrt{1 - \kappa^2 \sin^2 \phi}\, \mathrm{d}\phi = \int_0^K (\mathrm{dn}^2 u)\, \mathrm{d}u = E(\kappa, \pi). \tag{1.28}$$

The length of the rope can be expressed with $E(\kappa, \pi)$ and $2K(\kappa)$ as

$$L(\kappa) = \frac{4cE(\kappa)}{1 - \kappa^2} - 2a = \frac{4aE(\kappa)}{(1 - \kappa^2)K(\kappa)} - 2a.$$

1.5 Variation principle

We discussed the rotating rope as shape balanced with the centrifugal force, while the hanging rope was with gravity. In this section, we show that this problem can be treated by the *variation principle* for the results.

Using this principle, the problem is formulated for the potential energy of the rope, i.e.

$$U = \int_0^L \rho y'^2 \mathrm{d}s = \rho \int_{+a}^{-a} y'^2 \sqrt{1 + \left(\frac{\mathrm{d}y'}{\mathrm{d}z}\right)^2}\, dz,$$

to be minimized under a constant length $L = \int_{-a}^{+a} \sqrt{1 + \left(\frac{\mathrm{d}y'}{\mathrm{d}z}\right)^2}\, \mathrm{d}z = \text{const.}$

The method is applied for a new function $F = U/\rho - \alpha L$, where α is a parameter to minimize F, arriving at

$$\delta F = \int_{z_1}^{z_2} \left\{ \frac{\partial F}{\partial y'} - \frac{\mathrm{d}}{\mathrm{d}z} \frac{\partial F}{\partial(\mathrm{d}y'/\mathrm{d}z)} \right\} \delta y' \mathrm{d}z = 0.$$

Hence, we have Euler's equation

$$\frac{\mathrm{d}}{dz} \frac{\partial F}{\partial(\mathrm{d}y'/\mathrm{d}z)} - \frac{\partial F}{\partial y'} = 0,$$

which can be written explicitly as $\frac{\mathrm{d}}{\mathrm{d}z}\left\{ \frac{(y'^2 - \alpha\rho)\frac{\mathrm{d}y'}{\mathrm{d}z}}{\sqrt{1 + \left(\frac{\mathrm{d}y'}{\mathrm{d}z}\right)^2}} \right\} - 2y' \sqrt{1 + \left(\frac{\mathrm{d}y'}{\mathrm{d}z}\right)^2} = 0.$

Manipulating and integrating after, we arrive at $\frac{1}{2}\ln\left\{1+\left(\frac{dy'}{dz}\right)^2\right\} = \ln(\alpha\rho - y'^2) +$ const., where $y' = 0$ when $dy'/dz = 0$. Therefore,

$$\left(\frac{dy'}{dz}\right)^2 = \frac{2(b^2 - y'^2)}{\alpha\rho - b^2}\left\{1 + \frac{b^2 - y'^2}{2(\alpha\rho - b^2)}\right\},$$

giving the identical expression to (1.25), if $\frac{2}{\alpha\rho - b^2} = \frac{\rho\omega^2}{T_0}$.

For a hanging rope, we consider for the gravitational potential $U = \int_0^l \rho g y \, ds$ to be minimized with a constant length $l = \int_0^l ds$. Therefore the variation principle needs to minimize the integral $\int_0^l (y + \alpha\rho g)\sqrt{1 + \left(\frac{dy}{dz}\right)^2}\, dz$, leading to the relation $\left(\frac{dy}{dz}\right)^2 = \beta(y + \alpha\rho g)^2$ and $\beta = \frac{1}{(\alpha\rho - b^2)^2}$, which can be solved as $y = a\cosh\frac{z}{a} - \alpha\rho g$.

1.6 Buckling of an elastic rod

Buckling is an example of compressed elastic rod, representing a basic engineering issue for structural stability. In the physics of nonlinear dynamics, it is a fundamental phenomena of bifurcation. In this section, a model of a compressed rod at both ends is discussed, as illustrated in figure 1.5.

When forces $\boldsymbol{F}$ and $-\boldsymbol{F}$ exerted at two ends are sufficiently weak, the rod is just compressed with no deformation, but if the forces exceed the critical strength F_0 the

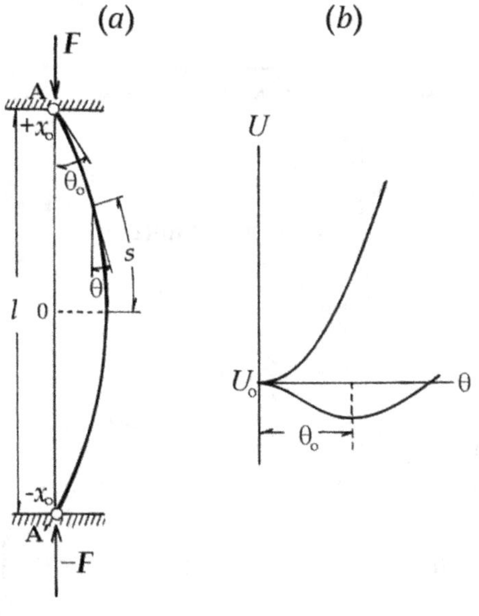

Figure 1.5. (*a*) Model for a buckled rod. (*b*) Energies of unbuckled rod U_0 compared with buckled rod $U(\theta_0) < U_0$.

rod is bent as illustrated in the figure. Normally, such a distortion is bifurcate for $F > F_o$, depending on the shape of the cross-sectional area.

For a rod bent as described by the angle $d\theta$ and radius R of curvature as in figure 1.5, we have the relation

$$\frac{1}{R} = -\frac{d\theta}{ds} = \frac{F}{B}y,$$

where B is a constant of the rod, i.e. material and cross-sectional area. Since we have a relation $\frac{dy}{ds} = \sin\theta$ in the figure, the equation

$$\frac{d^2\theta}{ds^2} = -c^2 \sin\theta \quad \text{where } c^2 = \frac{F}{B} \tag{1.29}$$

can be written for the model of buckling.

We consider that the rod is attached at points A and A' with no distortion, where $R = \infty$ signifies discontinuity of curvature at both ends. The attachment detail should be attributed to the physical properties of the rod and strength of $\boldsymbol{F}$, however we must accept such simplification for convenience of macroscopic description.

Specifying A and A' by coordinates x_1 and x_2, respectively, the distance AA' = $2l$ is variable with applied force F, and we calculate s from the center along the curved rod.

Referring to (1.1) and (1.29), we write $\sin\frac{\theta}{2} = \kappa \sin\varphi$ to obtain

$$\int_0^{\varphi} \frac{d\varphi}{\sqrt{1 - \kappa^2 \sin^2\varphi}} = cs \qquad \text{or} \qquad \varphi = \mathrm{am}(cs)$$

and

$$\cos\frac{\theta}{2} = \sqrt{1 - \kappa^2 \mathrm{sn}^2(cs)} = \mathrm{dn}(cs).$$

Hence

$$\frac{dy}{ds} = 2\kappa\, \mathrm{sn}(cs)\, \mathrm{dn}(cs),$$

from which we derive

$$y = -\frac{2\kappa}{c}\mathrm{cn}(cs), \tag{1.30}$$

where the constant for integration is zero.

For a small κ, the above integrand can be expanded as

$$K(\kappa) = \int_0^{\frac{\pi}{2}} \left(1 + \frac{\kappa^2}{2}\sin^2\varphi + ...\right) d\varphi = \frac{\pi}{2}\left(1 + \frac{\kappa^2}{4} + ...\right) = cl = \sqrt{\frac{F_o}{B}}\, l = \frac{\pi}{2}.$$

In this case, the rod is buckled with $F > F_o$, otherwise remaining straight for $F < F_o$.

The buckling energy composed of strain potential due to applied forces $\pm\mathbf{F}$ is expressed by

$$U = \int_0^l \left\{ \frac{B}{2}\left(\frac{\mathrm{d}\theta}{\mathrm{d}s}\right)^2 + F\cos\theta \right\} \mathrm{d}s = \int_0^l V(s)\,\mathrm{d}s,$$

where $V(s)$ is the energy density per unit length. Minimizing U, Euler's equation can be written as in section 1.5, from which the energy variation is expressed as

$$U - U_0 = 2\kappa^2 F \int_0^l \left\{ \mathrm{cn}^2(cs) - \mathrm{sn}^2(cs) \right\} \mathrm{d}s \tag{1.31a}$$

where $U_0 = Fl$ represents the potential energy of the straight rod. Using relations $cl = K(\kappa)$ and $\sin\varphi = \mathrm{sn}\,u$, (1.31$a$) can be expressed as

$$U - U_0 = \frac{2\kappa^2 Fl}{K(\kappa)} \int_0^{K(\kappa)} (\mathrm{cn}^2 u - \mathrm{sn}^2 u)\,\mathrm{d}u. \tag{1.31b}$$

Manipulating further with the formula

$$E(\kappa, \phi) = \int_0^u (\mathrm{dn}^2 u)\,\mathrm{d}u = \int_0^u (1 - \kappa^2 \mathrm{sn}^2 u)\,\mathrm{d}u,$$

and setting $u = K(\kappa)$ and $\sin\phi = \mathrm{sn}\,u$, we obtain $E(\kappa) = K(\kappa) - \kappa^2 \int_0^{K(\kappa)} (\mathrm{sn}^2 u)\mathrm{d}u$ and

$$U - U_0 = \frac{4Fl}{K(\kappa)} \left\{ E(\kappa) - \left(1 - \frac{\kappa^2}{2}\right) K(\kappa) \right\}. \tag{1.32c}$$

From formulas $K(\kappa) = \frac{\pi}{2}\left(1 + \frac{\kappa^2}{4} + \ldots..\right)$ and $E(\kappa) = \frac{\pi}{2}\left(1 - \frac{\kappa^2}{4} - \ldots..\right)$, we notice that $U < U_0$ for $\kappa \neq 0$, implying that the buckled rod has always a lower energy than the straight one. Therefore, we may consider that $\kappa \neq 0$ is a parameter for a strained rod.

For a small κ, the stress angle can be considered to be infinitesimal $\mathrm{d}\theta$, but the cross-sectional area is represented by the thickness $\mathrm{d}s$. The strain energy per unit length can then be expressed by $\frac{1}{2}\frac{\alpha\theta^2}{2l}$, where α is a constant. Referring to figure 1.5, the potential energy for bending is expressed as $2(R\sin\theta - l)F$, where $R\theta = 2l$.

Expanding $\sin\theta$, for a small κ, we have

$$U = \left(\frac{\alpha}{2l} - \frac{Fl}{24}\right)\theta^2 + \frac{Fl}{5!}\theta^4 - \ldots..,$$

indicating that U = minimum at $\theta = 0$ for $F < F_0 = \frac{12\alpha}{l^2}$. If $F > F_0$, from $\frac{\partial U}{\partial\theta} = 0$ the angle θ_0 can be determined by

$$\left(\frac{\alpha}{2l} - \frac{Fl}{24}\right) + \frac{F_0 l}{24}\theta_{F_0}^2 = 0 \quad \text{and} \quad \theta_{F_0} \propto \sqrt{F - F_0}.$$

It is noted that equation (1.29) can be linearized for $F < F_0$, and the buckling is attributed to the linear equation for $F > F_0$.

Exercise

1. Reviewing classical problems in this chapter regarding the total mechanical energy E, construct 'energy levels' to indicate the presence of 'forbidden energy bands'.
   'Energy gaps' characterize nonlinear dynamics, as will be discussed in later chapters.
2. The *least action principle* is normally employed for minimizing Gibbs' free energy in thermodynamics. We called the mathematical process *canonical transformation*. Is it a scientifically logical process? Discuss the process, in spite of the not quite scientific matter.

References

[1] Wikipedia http://en.wikipedia.org/wiki/Hyperbolic-function, http://en.wikipedia.org/wiki/Elliptic-function, http://en.wikipedia.org/wiki/Elliptic-integral

[2] Bowman F 1961 *Introduction to Elliptic Functions with Applications* (New York: Dover)
Greenhill A G 1959 *The Application of Elliptic Functions* (New York: Dover)
Hancock H 1958 *Lecture on Theory of Elliptic Functions* (New York: Dover)

IOP Publishing

Introduction to the Mathematical Physics of Nonlinear Waves
(Second Edition)

Minoru Fujimoto

Chapter 2

Wave propagation, singularities, and boundary conditions

In this chapter, the origin of nonlinear wave propagation in media is discussed, where mutual interactions, singularities and boundary conditions are responsible for deviation from linearity at low energies. Nevertheless, traditionally these interactions are considered as perturbations to linearity. Therefore, it is necessary to view the mathematical consequences from the principle of physics. Here, the specific property of Eckart potential is discussed, as it will be required in following chapters.

2.1 Elastic waves along a linear string in infinite length

2.1.1 Phase and amplitude of propagation

Basic excitations in a condensed matter exhibit propagating waves in one dimension in the x-direction. Considering a string of infinite length embedded in a medium, we describe the displacement $y = y(x, t)$ of a differential segment $\mathrm{d}s$ of mass density ρ, i.e. the mass is $\rho \mathrm{d}s$. The wave is described by the equation

$$\frac{\partial^2 y}{\partial x^2} - \frac{1}{v^2}\frac{\partial^2 y}{\partial t^2} = 0. \tag{2.1}$$

Regarding the space–time coordinates x and t along the string, v is the speed of propagation determined by the tension T_s on the string in the density $\rho(x, t)$. The general solution of (2.1) can be expressed as

$$y = f(x - vt) + g(x + vt). \tag{2.2}$$

Representing waves propagating in opposite directions, $\pm(x \mp vt)$. Analyzing by *Fourier's expansion*, (2.2) can be written as

$$y = \frac{1}{2\pi}\int_{-\infty}^{+\infty} Y(\omega)e^{i(kx-\omega t)}\mathrm{d}\omega, \tag{2.3}$$

where $Y^*(-\omega) = Y(+\omega)$ is called the Fourier amplitude, and k the wavevector for the relation $\omega = vk$.

The phase variable $\phi = kx - \omega t$ in (2.3) is normally employed in the one-cycle range $0 \leqslant \phi \leqslant 2\pi$ for convenience, while the Fourier amplitude is significant for the energy transport. Also, the space–time transformation can be discussed in terms of the phase ϕ. If there is no mechanism for energy dissipation, the phase ϕ determines linear propagation described by (2.1).

2.1.2 Energy flow

Physically, waves represent the excitation energy transported through the medium. In a range between x_1 and x_2, we can define kinetic and potential energies, K and v, respectively, composing vibrational energy $E = K + V$ per unit length of the string. Here, arising from the speed $\dot{y}(x, t)$ and the displacement $\mathrm{d}y(x, t)$, respectively, K and V, are expressed as

$$K = \frac{1}{2}\int_{x_1}^{x_2} \rho\left(\frac{\partial y}{\partial t}\right)^2 \mathrm{d}x \quad \text{and} \quad V = T_s\int_{x_1}^{x_2}\left(\sqrt{1 + \left(\frac{\partial y}{\partial x}\right)^2} - 1\right)\mathrm{d}x \approx \frac{T_s}{2}\int_{x_1}^{x_2}\left(\frac{\partial y}{\partial x}\right)^2 \mathrm{d}x,$$

where the latter represents the balanced tensions between displaced segments $\mathrm{d}x = x_2 - x_1$ of the string. Therefore, for small vibrational amplitudes, E can be written as

$$E = \frac{\rho}{2}\int_{x_1}^{x_2}\left\{\left(\frac{\partial y}{\partial t}\right)^2 + v^2\left(\frac{\partial y}{\partial x}\right)^2\right\}\mathrm{d}x \quad \text{and} \quad v^2 = \frac{T_s}{\rho}.$$

Hence, we obtain

$$\frac{\mathrm{d}E}{\mathrm{d}t} = \rho\int_{x_1}^{x_2}\left(\frac{\partial y}{\partial t}\frac{\partial^2 y}{\partial t^2} + v^2\frac{\partial y}{\partial x}\frac{\partial^2 y}{\partial x^2}\right)\mathrm{d}x \approx \rho v^2\int_{x_1}^{x_2}\frac{\partial}{\partial x}\left(\frac{\partial y}{\partial t}\frac{\partial y}{\partial x}\right)\mathrm{d}x.$$

Defining the energy density $\rho_E(x, t)$ and corresponding current density $J(x, t)$ from $E = \int_{x_1}^{x_2}\rho_E \mathrm{d}x$ and $J(x, t) = -\rho v^2 \frac{\partial y}{\partial t}\frac{\partial y}{\partial x}$, respectively, the continuity relation of energy-current can be expressed as

$$\frac{\mathrm{d}E}{\mathrm{d}t} = \frac{\mathrm{d}}{\mathrm{d}t}\int_{x_1}^{x_2}\rho_E \mathrm{d}x = -\int_{x_1}^{x_2}\frac{\partial J(x, t)}{\partial x}\mathrm{d}x = -J_{x_1}(t) + J_{x_2}(t), \tag{2.4a}$$

namely

$$\frac{\partial \rho_E}{\partial t} + \frac{\partial J}{\partial x} = 0. \tag{2.4b}$$

The current density plays a significant role at *singularities* in the media. There are several types of singularities that exist in the media, as will be discussed later. According to (2.4*b*), if the media are characterized by electric conductivity, the energy can be transferred to heat as an example, causing scattering phenomena. Otherwise $\Delta J = 0$ will not interfere wave propagation.

2.1.3 Scattering by an oscillator

Figure 2.1 shows an elastic string connected with a vertical spring at the central point M, while the two ends are connected at fixed points, A and B. We consider this arrangement as a model of singularity for wave propagation at the point M.

The equation of motion of mass m for the vertical spring at M can be written as

$$m\frac{d^2y}{dt^2} = -K_o y + T_s\left\{\left(\frac{\partial y_>}{\partial x}\right)_{x=0} - \left(\frac{\partial y_<}{\partial x}\right)_{x=0}\right\} e^{-i\omega t}, \tag{2.5}$$

where K_o and ω are the restoring constant and frequency of the spring, respectively. We solve (2.5) for the transversal displacement $y = y(0, k)e^{-ikvt}$, where $\omega = vk$.

Writing

$$y_>(x, k) = e^{ikx} + R(k)e^{-ikx} \quad \text{and} \quad y_<(x, k) = T(k)e^{ikx} \tag{2.6}$$

to set the boundary condition $y(0, k) = y_>(0, k) = y_<(0, k)$, we obtain the relation

$$y(0, k) = \frac{2i\kappa k}{k^2 - k_o^2 + 2i\kappa k}, \tag{2.7}$$

where $k_o v = \sqrt{K_o/m}$ is a resonant frequency, and $\kappa\, v = \sqrt{T_s/m}$. In addition, we have the transmission and reflection constants

$$T(k) = y(0, k) \quad \text{and} \quad R(k) = T(k) - 1 = \frac{k^2 - k_o^2}{k^2 - k_o^2 + 2i\kappa k},$$

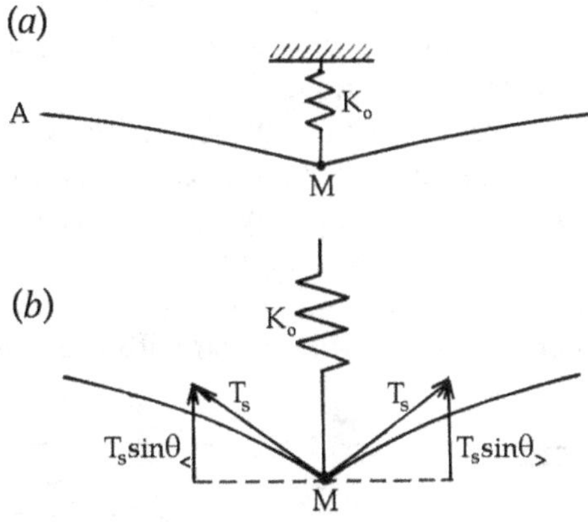

Figure 2.1. (*a*) A long elastic string held horizontally with a vertical spring attached at M. The spring constant is K_0. (*b*) Tensions T_s balanced with the spring force K_0.

respectively. For $k = k_o$, in particular, $T(k) = 1$ and $R(k) = 0$, indicating that the wave is resonant and there is no reflection. If $k \neq k_o$ on the other hand, the spring shows a singularity as determined by $k^2 - k_o^2 + 2i\kappa k = 0$, i.e.

$$k = -i\kappa \pm \sqrt{k_o^2 - \kappa^2}, \tag{2.8}$$

showing two poles of the function $y(0, k)$ near the imaginary axis, if $k_o \approx \kappa$.

Therefore, a force exists at a singularity that can be expressed as

$$\mathrm{T}_s\left\{\left(\frac{\mathrm{d}y_>}{\mathrm{d}x}\right)_{x=0} - \left(\frac{\mathrm{d}y_<}{\mathrm{d}x}\right)_{x=0}\right\}e^{-i\omega t} = ik\mathrm{T}_s\{2T(k) - 1\}e^{-i\omega t},$$

where the second term on the right, $\frac{\mathrm{d}y_>}{\mathrm{d}x} = -i\omega\, T(k)$, may be considered for an external force applied to the right side. Modifying (2.5) as

$$\frac{\mathrm{d}^2 y_>}{\mathrm{d}t^2} + 2\kappa v\frac{\mathrm{d}y_>}{\mathrm{d}t} + \kappa^2 v^2 y_> = \mp i\mathrm{T}_s\sqrt{k_o^2 - \kappa^2}\; e^{-i\omega t},$$

where the second term on the left represents *damping* of propagating energy. Normally, such damping occurs in the flowing energy toward the surrounding medium, establishing thermodynamic equilibrium. It is significant that the force on the right side is responsible for *dispersion* of frequency.

Such a singularity is generally dissipative as well as dispersive in character. Further, notable for resonance is that at $k_o = \kappa$, both phases of $y_>$ and $y_<$ are identical to $\phi = \kappa x - \omega t$, corresponding to $R = 0$ and $T = 1$.

2.2 Microwave transmission

Electromagnetic waves between parallel conducting lines (or plates) are a well-studied model for practical energy transmission known as a coaxial cable or waveguide [1]. Such guided waves are transmitted zigzag between parallel conductors, traveling as a group of plane waves at a *group velocity* v_g *of two components* along the direction of the waveguide. Thus, the guided wave at a frequency ω is signified by a wavevector $k = \omega/c$, where c is the speed of propagation.

Figure 2.2 shows such a waveguide as simplified by parallel lines, confining microwaves between the two. Here, two regions connecting with a *resonator* are marked *in and out*, representing the singularity of propagation. The waves reflect backward, transmitting through, when coming in from the left and going out to the right. Ignoring the detail of the resonator, we can discuss the scattering of plane waves of the propagating energy in one dimension with the figure.

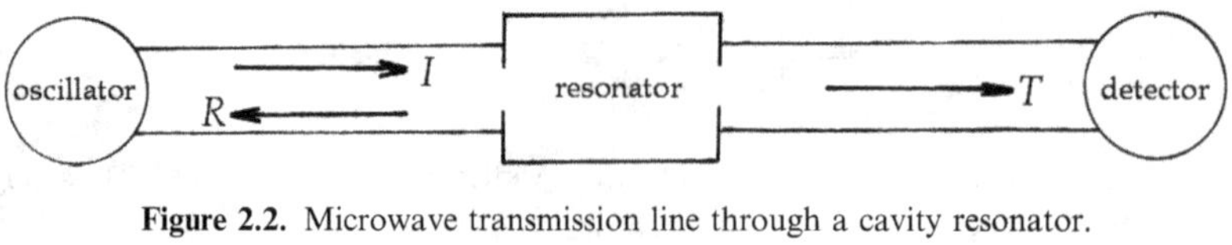

Figure 2.2. Microwave transmission line through a cavity resonator.

First, denoting the gap between parallel lines as $2d$, we write the wave equation

$$\frac{\partial^2 y}{\partial t^2} - v_g^2 \frac{\partial^2 y}{\partial x^2} = 0 \qquad \text{where} \quad \frac{\omega^2}{v_g^2} = k^2 - \frac{1}{4d^2} \tag{2.9}$$

for this waveguide. Noted from the last relation in (2.9), the frequency ω is restricted to $k > \frac{1}{2d}$ for the wave to propagate, namely $2d = \lambda_g$ is called the cut-off wavelength.

Letting $y = y(x)e^{-i\omega t}$, we obtain the equation

$$\frac{d^2 y}{dx^2} + k^2 y = K(x)y \qquad \text{where} \quad K(x) = \frac{1}{4d^2}.$$

Here, $K(x)$ behaves like a spring constant, signifying the electromagnetic energy is confined to the resonator, allowing to consider that the restoring energy $-\frac{1}{2}K(x)y^2$ is responsible for resonance. In the region $\Delta x_o = \pm\, n\lambda_g/2$ (n: integer) of the resonator, a large electromagnetic energy is stored at resonance.

The scattering process can then be discussed for the resonator to represent a singularity of nonlinear waves, illustrated in figure 2.2. For the left side, resonator and right side, respectively, we write

$$y_<(x) = e^{ikx} + R\, e^{-ikx},$$

$$y(x) = A\, e^{ik'x} + B\, e^{-ik'x} \qquad \text{where} \quad k' = \omega/v_g$$

and

$$y_>(x) = T\, e^{-ikx}.$$

At the resonator, we have boundary conditions

$$y_>(0) = y(0) \qquad \left(\frac{dy_>}{dx}\right)_{x=0} = \left(\frac{dy}{dx}\right)_{x=0} \qquad \text{at} \quad x = 0,$$

and

$$y(x_o) = y_<(x_o) \qquad \left(\frac{dy}{dx}\right)_{x=x_o} = \left(\frac{dy_<}{dx}\right)_{x=x_o} \qquad \text{at} \quad x = x_o,$$

respectively. Hence, we obtain

$$1 + R = A + B \qquad \text{and} \qquad ik(1 - R) = ik'(A - B) \quad \text{at} \quad x = 0$$

and

$$A\, e^{ik'x_o} + B\, e^{-ik'x_o} = T\, e^{ikx_o} \qquad ik'(A\, e^{ik'x_o} - B\, e^{-ik'x_o}) = ikT\, e^{ikx_o} \quad \text{at} \quad x = x_o.$$

Solving these equations for R and T at $k = k'$, we obtain $R = B$ at $x = 0$ and $T = A$ at $x = x_o$.

In microwave electronics, such resonators are fabricated for practical use, and the reflection and transmission are detected by observing resonance phenomena. In contrast, the observed $|R|^2$ and $|T|^2$ contain information of the resonator related to two components A and B, serving as study objectives in microwave spectroscopy.

2.3 Wave equations

2.3.1 Schrödinger's equation

In wave mechanics, scatterings of an electron by an atom is discussed by Schrödinger's equation $-\frac{\hbar^2}{2m}\frac{d^2\psi}{dx^2} + V(x)\psi = E\psi$. Letting $\frac{2mE}{\hbar^2} = k^2$ and $2V(x)\hbar^2 = U(x)$, the equation can be expressed as

$$\frac{d^2\psi}{dx^2} + \{k^2 - U(x)\}\psi = 0 \tag{2.10}$$

in one dimension.

Judging from the atomic scale, the potential $U(x)$ is highly localized inside the medium, so that (2.10) is a useful equation to deal with scatterings. Considering a free electron described by $\psi(x) \propto e^{\pm ikx}$ at $x = +\infty$ or $x = -\infty$, we can discuss its scattering by the target atom, keeping the concepts of reflection R and transmission T in mind. In this case, writing

$$\psi_< = A\, e^{ikx} + AR\, e^{-ikx} \quad \text{and} \quad \psi_> = AT\, e^{ikx}$$

for $U(x) = U_0\delta(x)$, $U_0 = 2mV_0/\hbar^2$ and $\delta(x) = \int_{-a}^{+a} dx$, we can express

$$R = \frac{-i\, U_0 a}{k + i\, U_0 a} \quad \text{and} \quad T = \frac{k}{k + i\, U_0 a}.$$

Accordingly, if $k = -i\, U_0 a$ and $U_0 < 0$, we have $R(\pm a) = \mp\infty$ and $T(\pm a) = \pm\infty$, although (2.11) indicates a particle trapped within $-a < x < a$ of the potential $U(x)$.

However, in a previous significant investigation in early physics a wave $\psi = e^{\pm kx} \sim \cosh x$ leading to $U(x) \sim 2\text{sech}^2 x$ was studied between $-i\, U_0 a < x < +i\, U_0 a$. In this case, the wave equation can be expressed as

$$\frac{d^2\psi}{dx^2} + \{-k^2 + U(x)\}\,\psi = 0, \tag{2.11}$$

corresponding to the eigenvalue $-k^2$.

2.3.2 Two-dimensional free waves in heterogeneous water

It is significant to see that such a potential $U(x) \sim -2\,\text{sech}^2 x$ exists in nature. Actually, the potential can be considered for the wave in the field of gravity, confirming for $U(x)$ to be like an *isolated pulse*, which will be verified later as real.

Figure 2.3 shows deep water extending in the xz-coordinate plane. In the presence of a gravitational field, the density function is expressed as $\rho(x, z)$, considering the gravity as uniform along the z-axis.

With such a density function $\rho(z)$, the water can be regarded to be of infinite extent in depth z, whereas the sound pressure $p(x, z)$ can penetrate along the z-direction, disregarding the y-direction. We therefore write the wave equation as

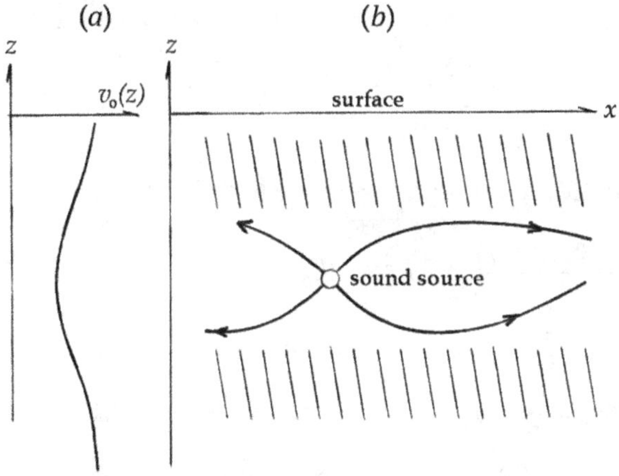

Figure 2.3. Sound propagation under water of a variable vertical density. (*a*) The speed of propagation $v_0(z)$ vs depth z. (*b*) Horizontal layer for sound propagation.

$$\frac{\partial^2 p}{\partial x^2} + \frac{\partial^2 p}{\partial z^2} + \frac{\omega^2}{v(z)^2} p = 0, \tag{2.12}$$

where the sound velocity $v(z)$ is a function of the depth z, and expressed from the relation $v(z)^2 = \frac{v_0^2}{1 + \mu(z)^2}$; here $\mu(z)$ is another function.

Setting $p(x, z) = A(z)e^{ikx}$ in (2.12), we obtain

$$V(z) = -\frac{\omega^2 \mu(z)^2}{v_0^2} \quad \text{and} \quad k_0 = \frac{\omega}{v_0}.$$

In the presence of a sound source of intensity Q located at a point (x, z), $V(z)$ can be modified by Q. Assuming $\mu(k_0 z) = 2\,\text{sech}(k_0 z)$, as inferred from the potential U_0, (2.12) can be written as

$$\frac{\partial^2 p}{\partial x^2} + \frac{\partial^2 p}{\partial z^2} + k_0^2\{1 + 2\,\text{sech}^2(k_0 z)\}p = -Q\,\delta(x)\,\delta(z), \tag{2.13}$$

where the potential $U(x, 0) = -2k_0^2\,\text{sech}^2(k_0 x)$ is permitted.

It is noted that (2.13) satisfies the boundary conditions for $p = 0$ at $(x, +\infty)$. However, k_0 is distributed in practice, requiring applied pressure to be redefined. Hence, the pressure Fourier transform $\bar{p}(k_x, z)$ of pressure $p(z)$ should be discussed with its Fourier transform $\bar{p}(k_x)$ given by

$$p(x) = \frac{1}{2\pi}\int_{-\infty}^{+\infty} \bar{p}(k_x)e^{ik_x x}dk_x \quad \text{and} \quad \delta(x) = \int_{-\infty}^{+\infty} e^{ik_x x}dk_x.$$

We therefore write the equation for $\bar{p}(k_x)$ as

$$\frac{d^2\bar{p}}{dz^2} + \left(k_0^2 - k_x^2 + 2k_0^2\,\text{sech}^2 k_0 z\right)\bar{p} = -Q(z) \quad \text{at} \quad x = 0 \tag{2.14a}$$

from which pressures $\bar{p}_>$ and $\bar{p}_<$ at both sides of the source are related by

$$\left(\frac{\mathrm{d}\bar{p}_>}{\mathrm{d}z}\right)_{x=0} - \left(\frac{\mathrm{d}\bar{p}_<}{\mathrm{d}z}\right)_{x=0} = -Q \quad \text{at} \quad x = 0. \tag{2.14b}$$

On the other hand, for $x > 0$ and $x < 0$, we have homogeneous equations

$$\frac{\mathrm{d}^2 p_{>,<}}{\mathrm{d}z^2} + \left(k_\mathrm{o}^2 - k_x^2 + 2k_\mathrm{o}^2 \operatorname{sech}^2 k_\mathrm{o} z\right) p_{>,<} = 0. \tag{2.15}$$

Defining $k_\mathrm{o}^2 - k_x^2 = k_z^2$, we obtain the solution of (2.15) as

$$\bar{p}_> = A(k_x)e^{+ik_z z}(ik_z - k_\mathrm{o} \tanh k_\mathrm{o} z) \text{ and } \bar{p}_< = A(k_x)e^{-ik_z z}(ik_z + k_\mathrm{o} \tanh k_\mathrm{o} z), \tag{2.16}$$

with the same factor $A(k_x)$. At $z = 0$, we have $p_> = p_<$, so that

$$A(k_x) = \frac{Q}{2(2k_\mathrm{o}^2 - k_x^2)}. \tag{2.17}$$

Therefore, the two-dimensional field of sound is given by

$$p(x, z) = \frac{Q}{4\pi}\int_{-\infty}^{+\infty} \mathrm{d}k_x e^{i(k_x x + k_z|z|)} \frac{ik_z - k_\mathrm{o} \tanh k_\mathrm{o}|z|}{2k_\mathrm{o}^2 - k_x^2}, \tag{2.18}$$

which is composed of two independent orthogonal waves.

Such a complex integral in (2.18) can be evaluated with *Cauchy's theorem* of complex functions along a circular path for integration in $\boldsymbol{k}$-space as illustrated in figure 2.4, which are signified by two singular points at $k_x = \pm 2k_\mathrm{o}$. In the figure, these points are specified by a complex notation $k' + \mathrm{i}k'' = k_x + ik_z$, showing a broadened anomaly curve $k_x k_y = \text{const}$. Choosing a semi-circular path to avoid the singular pole and anomalies, the results of integration can be expressed by

$$p(x, z) = \frac{Q}{4\sqrt{2}} e^{i\left(\sqrt{2}k_x + \frac{1}{2}\right)} \operatorname{sech} k_o z + \frac{Q}{4\pi}\int dk_z e^{ik_z z}(ik_z - k_o \tanh k_o z)F(k_o), \tag{2.19a}$$

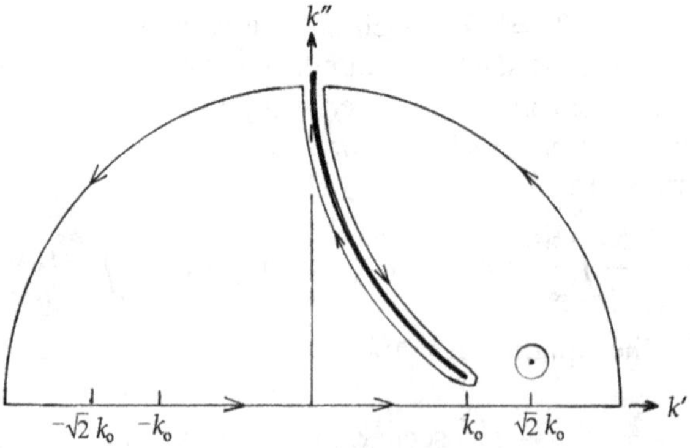

Figure 2.4. Cauchy's semi-circular diagram for the k-vector of sound propagation.

where

$$F(k_o) = \frac{k_o e^{ik_x\sqrt{k_o^2-k_z^2}}}{\sqrt{k_o^2 - k_z^2}(k_o^2 + k_z^2)}. \tag{2.19b}$$

It is noted that equation (2.19*b*) at the origin is given by a complex form, where the real and imaginary parts are mutually perpendicular. We can show that

$$\int_{-\infty}^{+\infty} (\text{sech}\, k_z z)\{ik_z - k_o(\tanh k_o z)\}\, e^{ik_o z} \mathrm{d}z = 0.$$

Indicating that the real and imaginary parts in (2.1*a*) are perpendicular at the origin.

2.3.3 Eckart's potential

Equation (2.11) is essential for nonlinear dynamics in the following chapters, so we discuss properties of the potential

$$V(x) = -V_o \text{sech}^2 \frac{x}{d} \qquad \text{where } V_o = \text{const.}, \tag{2.20a}$$

where the negative potential is characterized by the width $2d$, as shown in figure 2.5.

We rewrite the wave equation for convenience as

$$\frac{\mathrm{d}^2\psi}{\mathrm{d}x^2} + (\varepsilon + v\, \text{sech}^2 x)\psi = 0 \tag{2.20b}$$

where $\varepsilon = 2md^2E/\hbar^2$ and $v = 2md^2V_o/\hbar^2$. Following Morse and Feshbach [1], by changing $\psi(x)$ to $Y(u)$ as

$$\psi = A\, \text{sech}^\beta(x) Y(x) \quad \text{and then} \quad u = \frac{1}{2}(1 - \tanh x).$$

We can transform the above to a hypergeometric differential equation

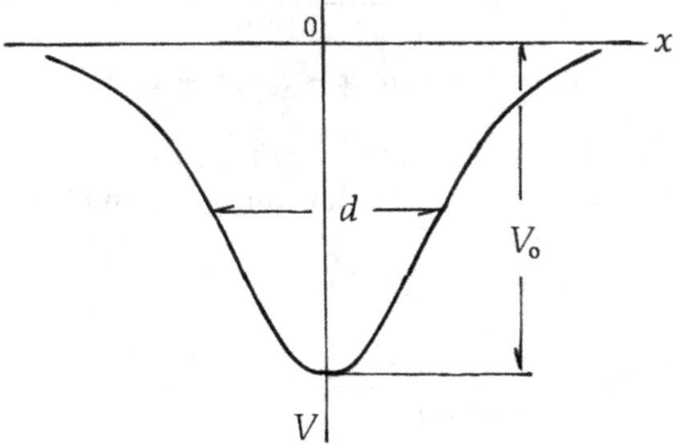

Figure 2.5. An Eckart's potential.

$$\frac{\mathrm{d}^2 Y}{\mathrm{d}x^2} - 2\beta(\tanh x)\frac{\mathrm{d}Y}{\mathrm{d}x} + (v - \beta^2 - \beta)(\mathrm{sech}^2\, x)Y = 0. \tag{2.21a}$$

This can then be re-expressed as

$$u(u-1)\frac{\mathrm{d}^2 Y}{du^2} + \{c - (a+b+1)u\}\frac{\mathrm{d}Y}{du} - ab\, u = 0, \tag{2.21b}$$

where

$$c = 1 + \beta, \quad a + b + 1 = 2(1 + \beta) \text{ and } ab = \beta^2 + \beta - v. \tag{2.21c}$$

Equation (2.21a) is the standard form of hypergeometric equation whose solution for $x \to \infty$ is written as

$$Y(u) = F(a, b, c; u) = 1 + \frac{ab}{c}u + \cdots.$$

By definition, $x \to \infty$ corresponds to $u \to 0$ and $Y \to 1$; hence $\psi_{x\to\infty} \to A\, 2^{\beta} e^{-\beta x}$. In order for $\psi_{x\to\infty}$ to represent e^{-ikx} for $\varepsilon = \frac{\hbar^2 k^2}{2m}$, we should have the relation $\beta = -ikd$.

Solving (2.21b), values of a, b and c are determined as

$$a = \frac{1}{2} - ikd + \sqrt{v + \frac{1}{4}}, \quad b = \frac{1}{2} - ikd - \sqrt{v + \frac{1}{4}} \text{ and } c = 1 - ikd. \tag{2.21d}$$

For $\psi_{x\to-\infty}(x)$, however, it is convenient to expand $F(a,\ b,\ c;\ u)$ at $u = 1$, and we obtain

$$\begin{aligned} F(a,\ b,\ c;\ u) &= \frac{\Gamma(c)\Gamma(c-a-b)}{\Gamma(c-a)\Gamma(c-b)}F(a,\ b,\ a+b-c+1;\ 1-u) \\ &\quad + (1-u)^{c-a-b}\frac{\Gamma(c)\Gamma(a+b-c)}{\Gamma(a)\Gamma(b)} \\ &\quad F(c-a,\ c-b,\ c-a-b+1;\ 1-u). \end{aligned} \tag{2.21e}$$

Noting that in (2.21e) the function $F(\ldots;1-u) \to 1$ when $u \to 1$, and $(1-u)^{c-a-b} = e^{-2\beta z}$. Characterizing the function $\psi_{z\to-\infty}$ in (2.20b) by the amplitude A, we have the relation $F(a,\ b,\ c;\ u-1) \to 1$, deriving $\mathrm{sech}^{\beta} z \approx 2^{\beta} e^{\beta z}$ and $\beta = -ikd$. We finally obtain the expression

$$\psi_{x\to-\infty} \approx A\, 2^{\beta}\left\{\frac{\Gamma(c)\Gamma(a+b-c)}{\Gamma(a)\Gamma(b)}e^{+ikx} + \frac{\Gamma(c)\Gamma(c-a-b)}{\Gamma(c-a)\Gamma(c-b)}e^{-ikx}\right\}.$$

At this point, it is interesting to notice that these coefficients have singularities due to the properties of Γ-functions, namely $\Gamma(a)\,\Gamma(b) \to \pm\infty$, if

$$a, \; b = -m = -1, -2, -3, \; \ldots . \tag{2.21f}$$

As seen from figure 2.6. In addition, from the relation

$$\Gamma(c-a)\,\Gamma(c-b) = \Gamma\left(\frac{1}{2} + \sqrt{v+\frac{1}{4}}\right)\Gamma\left(\frac{1}{2} - \sqrt{v+\frac{1}{4}}\right) = \frac{\pi}{\cos\left(\pi\sqrt{v+\frac{1}{4}}\right)},$$

becoming $\pm\,\infty$, when

$$\sqrt{v+\frac{1}{4}} = \frac{2n+1}{2} \qquad \text{for } n = 1, \; 2, \; \ldots,$$

i.e.

$$v = n(n+1). \tag{2.21g}$$

Under the condition of (2.21*g*), the Eckart potential exhibits no reflection as revised $\beta > 0$

$$\beta = n - m \quad \text{where } m = 0, \; 1, \; 2, \; \ldots, \; n-1 \tag{2.21h}$$

or define a new parameter, instead of β, as

$$\mathrm{p} = n, \; n-1, \; n-2, \; \ldots, \; 2, \; 1. \tag{2.21i}$$

Summarizing the above mathematics, the wave equation for the potential $V(x)$ in (2.20*a*) can be expressed in terms of p and n as in

$$\frac{\mathrm{d}^2\psi_{\mathrm{p},n}}{\mathrm{d}x^2} + \{-\mathrm{p}^2 + n(n+1)\,\mathrm{sech}^2 x\}\psi_{p,x} = 0. \tag{2.21j}$$

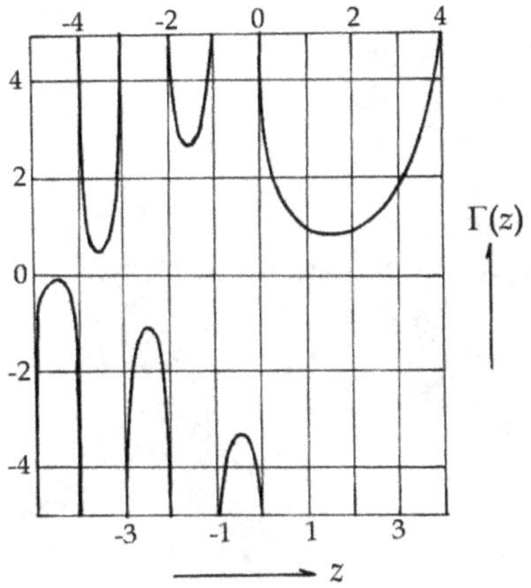

Figure 2.6. Mathematical properties of the Gamma function.

Though just mathematical, the above arguments show the potential in (2.20*a*) is real to be found in nature, as discussed later by soliton theory. Also, it is significant to discuss the reason why such potentials are discrete.

2.4 Sound propagation in air

In the classical viewpoint, the sound wave is generated by a momentum change caused by the pressure variation, which can be described by the equation

$$\frac{\partial v}{\partial t} + v\frac{\partial v}{\partial z} = -\frac{1}{\rho_0}\frac{\Delta p}{\Delta z}. \tag{2.22a}$$

Considering that air mass $\rho_0(Al) = m$ is determined by the static density ρ_0 and cylindrical volume Al of area A and height $\Delta z = l$. The pressure $p(z)$ is a function of z, and the still pressure is given by $p = p_0$.

The wave is not restricted to the vertical z-axis, but spread out in the horizontal xy-plane as well, so we have $p(z) = p_0 + \Delta p$, and

$$\frac{\partial \rho_z}{\partial t} + \rho_0\frac{\partial v}{\partial z} = 0 \quad \text{and} \quad \frac{\partial \rho_x}{\partial t} = 0. \tag{2.22b}$$

Further, assuming for ρ_z to be within the narrow vicinity in the x-axis, we obtain the relation

$$\frac{\partial^2 \rho_z}{\partial t^2} + \alpha(\rho_z - \rho_0) = \Delta p \quad \text{where } \alpha = \text{const.} \tag{2.22c}$$

Equation (2.22*a*) is nonlinear in terms of $v\frac{\partial v}{\partial z}$ on the left, whereas (2.22*b*) and (2.22*c*) are linear. The density components ρ_x and ρ_z are both functions of x and z, but we assume $|\rho_x| = |\rho_z| = \rho_0$. Introducing a set of *reduced variables*

$$x' = \sqrt{\alpha}\, x, \qquad t' = \sqrt{\alpha}\, t, \qquad \rho' = \frac{\rho_{x,z} - \rho_0}{\rho_0} \quad \text{and} \quad \tau = \frac{1}{2}k^3 t',$$

we can write equations

$$\frac{\partial v}{\partial t'} + \frac{\partial p'}{\partial x'} = 0, \qquad \frac{\partial \rho'}{\partial t'} - \frac{\partial v}{\partial x'} = 0 \quad \text{and} \quad \frac{\partial^2 \rho'}{\partial t'^2} + \rho' = p'. \tag{2.22d}$$

Setting that ρ', v, p' are all proportional to $e^{i(kx-\omega t)}$, we obtain a relation among amplitudes. Then, the following expression is derived from (2.22*d*) that is

$$\omega^2 = \frac{k^2}{1+k^2} \quad \text{or} \quad \omega \approx k - \frac{1}{2}k^3$$

for small value of ω.

2.5 Asymptotic approximation in air space

Born and Oppenheimer [2] proposed the use of the asymptotic approximation for sound propagation in condensed matter, regarding the masses of constituent atoms and molecules. In their theory, they considered that constituents are identifiable up to the accuracy of adiabatic approximation, restricting discussions within their approximation.

Therefore, any material should be dealt with the theory within adiabatic accuracy. In fact, those concepts like, solitons, Eckart's potentials, and other concepts essential to nonlinear dynamics are logically acquired as facts of nature, consistent with adiabatic approximation.

Here, continuing this section, we show that an important theorem in nonlinear dynamics, the Korteweg–deVries equation, can be derived in the discussion of sound propagation.

Writing $e^{i(kx'-\omega t')} = e^{ik(x'-t')}e^{-\frac{1}{2}k^3t'}$ for convenience, we define $\xi = k(x' - t')$ and $\tau = \frac{1}{2}k^3t'$ to change coordinates $(x', t') \to (\xi, \tau)$ in the following transformations

$$\frac{\partial}{\partial x'} \to k\frac{\partial}{\partial \xi} \qquad \text{and} \qquad \frac{\partial}{\partial t'} = -k\frac{\partial}{\partial \xi} + \frac{1}{2}k^3\frac{\partial}{\partial \tau}.$$

Equations (2.22*a*) and (2.22*b*) can then be transformed from

$$\frac{\partial v}{\partial t'} + v\frac{\partial v}{\partial x'} + \frac{\partial p'}{\partial x'} = 0 \qquad \text{and} \qquad \frac{\partial p'}{\partial t'} + \frac{\partial v}{\partial x'} + \frac{\partial(\rho' v)}{\partial x'} = 0$$

to

$$-\frac{\partial v}{\partial \xi} + \frac{1}{2}k^2\frac{\partial v}{\partial \tau} + v\frac{\partial v}{\partial \xi} + \frac{\partial p'}{\partial \xi} = 0 \quad \text{and} \quad -\frac{\partial \rho'}{\partial \xi} + \frac{1}{2}k^2\frac{\partial p'}{\partial \tau} + \frac{\partial v}{\partial \xi} + \frac{\partial(\rho' v)}{\partial \xi} = 0. \qquad (2.22e)$$

The third equation in (2.22*d*) is transformed as

$$p' = \rho' + k^2\frac{\partial \rho'}{\partial \xi} - k^4\frac{\partial^2 \rho'}{\partial \xi \partial \tau} + k^6\frac{\partial^2 \rho'}{\partial \tau^2}. \qquad (2.22f)$$

Equations (2.22*a*) and (2.22*b*) can be solved with respect to k^2 in an *asymptotic approximation*, where variables ρ', $v - v_0$ and p' are attributed to emerging nonlinear quantities in media under critical conditions. Expanding in the power series of k^2, we write

$$\rho' = k^2\rho_1' + k^4\rho_2' + \cdots,$$
$$v - v_0 = k^2 v_1 + k^4 v_2 + \cdots$$
$$\text{and} \qquad p' = k^2 p_1' + k^4 p_2' + \cdots.$$

Substituting these to (2.22*e*) and (2.22*f*), which are evaluated by factors of k^2, k^4, ... separately. From terms of k^2, we obtain

$$-\frac{\partial \rho_1'}{\partial \xi} + \frac{\partial v}{\partial \xi} - 0, \quad -\frac{\partial v}{\partial \xi} + \frac{\partial p'}{\partial \xi} = 0 \ \text{ and } \ p_1' = \rho_2',$$

while integrating the first two relations, we have $p_1' = \rho_1' = v_1 + \varphi(\tau)$ where $\varphi(\tau)$ is an arbitrary function of τ.

From terms of k^4,

$$-\frac{\partial \rho_2'}{\partial \xi} + \frac{1}{2}\frac{\partial \rho_1'}{\partial \tau} + \frac{\partial v_2}{\partial \xi} + \frac{\partial(\rho_1' v_1)}{\partial \xi} = 0,$$

$$-\frac{\partial v_2}{\partial \xi} + \frac{1}{2}\frac{\partial v_1}{\partial \tau} + v_1\frac{\partial v_1}{\partial \xi} + \frac{\partial p_2'}{\partial \xi} = 0$$

and

$$p_2' = \rho_2' + \frac{\partial^2 \rho_1'}{\partial \xi^2}.$$

Using the last expression in the first and second, we can eliminate p_1', ρ_2', v_2 to obtain relations for p_1', ρ_1', v_1, which are

$$\frac{\partial v_1}{\partial \tau} + 3v_1\frac{\partial v_1}{\partial \xi} + \frac{\partial^2 v_1}{\partial \xi^2} + \varphi\frac{\partial v_1}{\partial \xi} + \frac{\partial \varphi}{\partial \tau} = 0$$

and

$$\frac{\partial \rho_1'}{\partial \tau} + 3\rho_1'\frac{\partial \rho'}{\partial \xi} + \frac{\partial^3 \rho'}{\partial \xi^3} - \varphi\frac{\partial \rho'}{\partial \xi} - \frac{\partial \varphi}{\partial \tau} = 0.$$

Choosing the function $\varphi(\tau)$ to satisfy $\varphi\frac{\partial(v_1, \rho_1')}{\partial \xi} + \frac{\partial \varphi}{\partial \tau} = 0$, we obtain the equation for ρ_1', v_1, p_1', representing the potential $V_1(\xi, \tau)$ in common equation

$$\frac{\partial V_1}{\partial \tau} + 6V_1\frac{\partial V_1}{\partial \xi} + \frac{\partial^3 V_1}{\partial \xi^3} = 0. \tag{2.23a}$$

Equation (2.23*a*) is called the *Korteweg–deVries equation*. It is noted that the third derivative on the left is responsible for dispersion to originate from k^3 in the asymptotic approximation, describing nonlinear properties of media. Also, the potential $V_1(\xi, \tau)$ is essential for development of nonlinearity to take place by the force $-\frac{\partial^3 V_1}{\partial \xi^3} = -\frac{\partial \Delta U_1}{\partial \xi}$ in thermodynamic processes. Therefore, we can write

$$\left(\frac{\mathrm{d}\rho}{\mathrm{d}\xi}\right)_{\xi_0} = \frac{\partial \rho}{\partial \tau} \pm v\frac{\partial \rho}{\partial \xi_0} = -\frac{\partial \Delta U_1}{\partial \xi_0}, \tag{2.23b}$$

with respect to ξ_0, where $6V_1 = v$.

In addition, representing ρ_1', v_1, p_1', the potential $V_1(\xi, \tau)$ is proportional to $\rho_1'(\xi, \tau)$, confirming that the Weiss field for ferromagnetic media is a valid postulate supported by the asymptotic approximation in k^2-accuracy.

Exercises

1. In section 2.2, we derived expressions for R and T for an idealized one-dimensional resonator. Confirm the relation $|R|^2 + |T|^2 = 1$ to be the characteristic conservative feature.
2. How can a system of many particles interacting with a local potential be characterized as conservative? Discuss this problem in one dimension.
3. Discuss the reason why Galilean invariance of space–time is significant for practical periodic systems.
4. In section 2.5, a space–time coordinate change $(x', t') \rightarrow (\xi, \tau)$ is performed to derive from the Korteweg–deVries equation. What is the significance of this transformation?
5. The Born–Oppenheimer approximation is expressed by an asymptotic expansion to the order of k^4, according to section 2.5. The expansion can be signified for the limit for constituent atoms/molecules to be identified in general. Discuss the issue for practical crystals, where the presence of Weiss' field is mandatory.

References

[1] Morse P M and Feshbach H 1953 *Methods of Theoretical Physics* (New York: McGraw-Hill)
[2] Born M and Huang K 1968 *Dynamical Theory of Crystal Lattices* (London: Oxford University Press)

IOP Publishing

Introduction to the Mathematical Physics of Nonlinear Waves
(Second Edition)

Minoru Fujimoto

Chapter 3

Order variables for structural phase transitions

Representing clustered pseudospins in crystals, order variables in the collective mode are responsible for the structural phase transitions and following mesoscopic disorder. In this chapter, thermodynamic internal energy needs to be formulated as correlation energy regarding interactions among order variables, although normally considered as a constant of temperature. Collective order variables called *solitons* are used for structural phase transitions, constituting the basic objective in nonlinear dynamics in crystals.

3.1 Symmetry group in crystals

The structure of a crystal is characterized in a symmetrical form, transformable from one type to another by geometrical operations, constituting a crystal group. Such an operation can be expressed by a matrix potential Q_{ij}, where the coordinate axes are indicated by i and j. Considering lattice vibrations, the stable form of a crystal is maintained at centers of vibration. We therefore consider Hamiltonian $\mathcal{H}_{ij}$ for vibrational kinetic energies to represent the lattice.

Then the Lagrangian $\mathcal{L} = \mathcal{H}_{ij} + Q_{ij}$ integrated along a path in the dynamical phase space must be minimized to obtain the Hamiltonian equation for a canonical transformation to obtain $\langle i|\mathcal{H} - Q|j\rangle = E_i\langle i|\delta|j\rangle$, where $\langle i|\delta|j\rangle$ is known as Kronecker's delta, meaning that

$$\mathcal{H}_{ii} = Q_{ii} = E_i,$$

indicating that both $\mathcal{H}_{ij}$ and Q_{ij} are specified by a common eigenvalue E_i; i.e. *conservative* states correspond to a symmetry potential that corresponds to temperature.

doi:10.1088/978-0-7503-3759-5ch3

3.2 Solitons and the Ising model for pseudospin correlations

3.2.1 Pseudospin correlations

Judging from diffused x-ray diffraction experiments, certain atoms are in motion in a stable structure of crystals. Figure 3.1 illustrates a unit-cell structure of *perovskites* $BaTiO_3$ and $SrTiO_3$, where Ti^{4+} ions at the center of tetrahedral complex TiO_6^{2-} exhibit diffused x-ray diffraction, indicating fluctuation at certain temperatures. Such motion can be discussed with a tunneling model, offering an idea of pseudospin.

However, assuming for such fluctuations to occur along a symmetry axis, the motion can take place with the lowest energy at a given temperature, and cannot be harmonic, for the stability at a lattice point is maintained by harmonic motion with infinitesimal amplitude. Therefore, we consider a *finite shift* for tunneling, as shown in figure 3.2, calling a *finite pseudospin as soliton.* Solitons are characterized by a specific wavevector $\boldsymbol{q}$ and frequency $\tilde{\omega}$ for the order variable $\boldsymbol{\sigma}_{\boldsymbol{q}}$ and $\boldsymbol{\sigma}_{-\boldsymbol{q}}$. It is noted however that $\boldsymbol{q}$ should have a mesoscopic magnitude for structural change characterized by a wavelength longer than the unit-cell.

First, we need to discuss the dynamic matter to describe *collective motion* in real crystals, which is assumed to be uniform space enclosed by surfaces, where the dynamics can be analyzed by Fourier's analysis in a *Brillouin zone*. Hence, in the

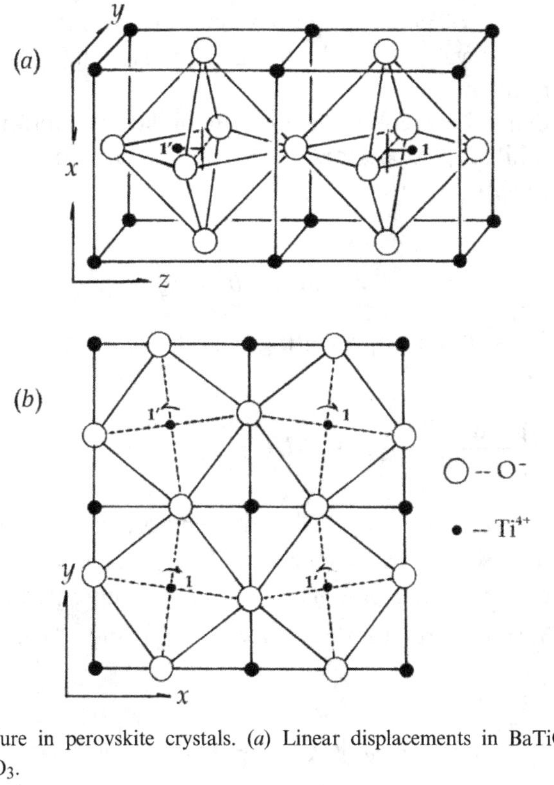

Figure 3.1. Cell structure in perovskite crystals. (*a*) Linear displacements in $BaTiO_3$ and (*b*) rotational displacements in $BaTiO_3$.

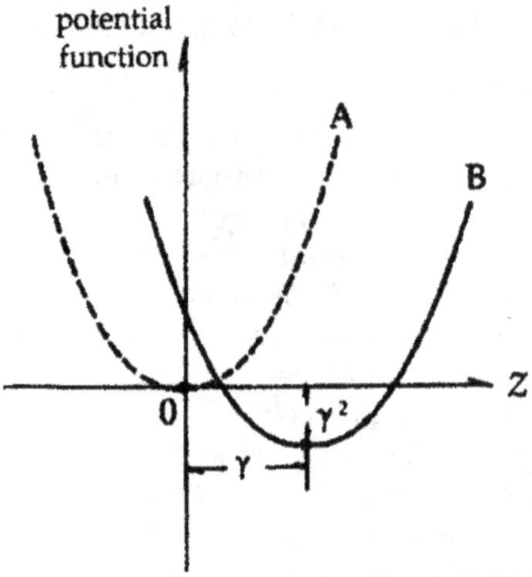

Figure 3.2. The potential energy curve for a finite displacement. A and B are the harmonic potentials $U(0)$ and $U(z)$ at 0 and z, respectively, which are shifted by a potential $z^2/2$. The parameter γ is a constant defined by (3.1).

following argument, all dynamical descriptions will be in a representative Brillouin zone for collective modes of order variables.

As illustrated in the figure, the pseudospin $\boldsymbol{\sigma}_q$ is characterized by the finite shift γ and negative depth $-\gamma^2$ of potential minimum, which are distinctive features from phonons. Such a model was previously discussed by Haken [1], hence our discussion is following his argument.

For phonons defined by collective motion of lattice vibrations, we have the creation and annihilation operators, b_q and $b_q^\dagger$, while for *solitons for specific molecular displacements* at $\boldsymbol{q}$, we also consider the corresponding operators defined by

$$b_q = \tilde{b}_q + \gamma \quad \text{and} \quad b_q^\dagger = \tilde{b}_q^\dagger + \gamma. \tag{3.1}$$

The anharmonic oscillator, representing $\boldsymbol{\sigma}_q$ and $\boldsymbol{\sigma}_{-q}$, can then be described by the equation of motion

$$\left\{-\frac{1}{2}\frac{\mathrm{d}^2}{\mathrm{d}t^2} + \frac{1}{2}z^2 - \Delta U(z)\right\}\psi_q(z) = \varepsilon_q\psi_q(z), \tag{3.2}$$

where $\varepsilon_q = E_q/\hbar\tilde{\omega}_q$ is an eigenvalue with respect to soliton energy, and $\Delta U = \gamma z$ corresponds to a mesoscopic displacement z. Here, $\psi_q(z)$ is the wave function of the oscillator characterized by inversion $\boldsymbol{q} \rightleftarrows -\boldsymbol{q}$ inside a finite periodic crystal.

For equation (3.2), we write that $z = \frac{1}{\sqrt{2}}\left(b_q + b_{-q}^\dagger\right)$ and $\frac{\mathrm{d}}{\mathrm{d}z}\left(b_q - b_{-q}^\dagger\right)$. Then, (3.2) can be expressed as

$$\left\{b_q^\dagger b_q - \gamma\left(b_q^\dagger + b_q\right)\right\}\psi_q = \left(\varepsilon_q - \frac{1}{2}\right)\psi_q. \tag{3.3}$$

However, we notice that the operators b_q and $b_q^\dagger$ do not constitute the commutation relation as boson particles. Therefore, leaving the relation $\left[b_q,\ b_q^\dagger\right] = 1$ to the case for $\gamma = 0$, the operators $\tilde{b}_q$ and $\tilde{b}_q^\dagger$ for a finite displacement $\gamma \neq 0$, i.e.

$$\left[\tilde{b}_q,\ \tilde{b}_q^\dagger\right] = \tilde{b}_q\tilde{b}_q^\dagger - \tilde{b}_q^\dagger b_q = 1 \tag{3.4}$$

can be assigned to soliton particles, obeying Bose–Einstein statistics.

Haken demonstrated in his book that $\psi\left(b_q,\ b_q^\dagger\right)$ can be transformed to $\tilde{\psi}\left(\tilde{b}_q,\ \tilde{b}_q^\dagger\right)$, thereby (3.3) is expressed as

$$\left(\tilde{b}_q\tilde{b}_q^\dagger - \gamma^2\right)\tilde{\psi}_q = \tilde{\varepsilon}_q\tilde{\psi}_q \quad \text{and} \quad \tilde{\varepsilon}_q = \varepsilon_q - \gamma^2 \quad \text{for } \gamma \neq 0, \tag{3.5}$$

which can alternatively be written as

$$\left(\tilde{b}_q^\dagger\tilde{b}_q\right)\tilde{\psi}_q = \lambda_q\tilde{\psi}_q \qquad \text{and} \qquad \lambda_q = \tilde{\varepsilon}_q - \frac{1}{2}. \tag{3.6}$$

Corresponding to the zero-point energy of phonons in crystal at $T = 0$ K, for λ_q in (3.6), no temperature is assigned yet, hence $\lambda_q = 0$ should be considered for a temperature under critical conditions. In addition, referring to a specific wavevector $\boldsymbol{q}$, the soliton should be a mobile particle like phonons in the momentum space, obeying Bose–Einstein statistics.

Now that operators $\tilde{b}_q^\dagger$ and $\tilde{b}_q$ are defined for inversion $\boldsymbol{q} \rightleftarrows -\boldsymbol{q}$, their interaction between $\boldsymbol{\sigma}_q$ and $\boldsymbol{\sigma}_{-q}$ is expressed as a scalar product $-J\ \boldsymbol{\sigma}_q \cdot \boldsymbol{\sigma}_{-q}$ for $J > 0$, referring to Bose–Einstein statistics. The probability to find a combination of pseudospins is given by

$$P = \frac{1}{2}(1 + \boldsymbol{\sigma}_q \cdot \boldsymbol{\sigma}_{-q}) \quad \text{for} \quad |\boldsymbol{\sigma}_{\pm q}| = 1, \tag{3.7}$$

so that for $\boldsymbol{\sigma}_{\pm q}$ in the molar unit, $\boldsymbol{\sigma}_q \cdot \boldsymbol{\sigma}_{-q} = \pm 1$ for parallel and antiparallel spins, giving rise to $P = 1, 0$, respectively, in thermal equilibrium. Namely, $\boldsymbol{\sigma}_q \| \boldsymbol{\sigma}_{-q}$ for $-J\ \boldsymbol{\sigma}_q \cdot \boldsymbol{\sigma}_{-q}$ is more stable than the antiparallel combination, as specified by symmetrical $\frac{\sigma_q + \sigma_{-q}}{\sqrt{2}}$.

3.2.2 Soliton correlations in pseudospin clusters

The second-order structural change is a major subject in irreversible thermodynamics of crystalline processes [1]. Among many types of crystal, perovskite crystals composed of pyramidal constituents provides a typical model for structural transitions. Figure 3.1 indicates that lattice structures of $BaTiO_3$ and$SrTiO_3$ composed of bipyramidal $(TiO_6)^{2-}$ complexes in the lattice, exhibiting an active role in the phase transition.

Characterized by a symmetry change from cubic to tetragonal, unit cells exhibit deformation at critical temperature as illustrated in the figure. While occurring in each unit cell after transition, the distortion linked with another cell, so they are grouped to form *clustered units* of small size. Hence, it is logical to consider the presence of clusters in the initial stage of structural change. Therefore, Landau [2] considered that the amplitude of $\boldsymbol{\sigma}_q$ changes as a function of time, which is equivalent for clusters to represent the order variable in transition processes.

In general, the correlation energy can be expressed by a Hamiltonian

$$\mathcal{H}_n = -\sum_m J_{mn}\boldsymbol{\sigma}_m \cdot \boldsymbol{\sigma}_n \tag{3.8}$$

to reach to new symmetry for a lower state. For this purpose, we write $\boldsymbol{\sigma}_m = \sigma_o \mathbf{e}_m$ and $\boldsymbol{\sigma}_n = \sigma_o \mathbf{e}_n$ for clusters *at m* and *n*, where σ_o is a constant amplitude, and $\mathbf{e}_m$ and $\mathbf{e}_n$ are unit vectors at sites m and n.

In the presence of a cluster conversion $\boldsymbol{q} \rightleftarrows -\boldsymbol{q}$ in one-dimension, we express

$$\mathbf{e}_{n\pm} = \mathbf{e}_q e^{i(\boldsymbol{q}\cdot\boldsymbol{r}_n - \tilde{\omega}t_n)} \pm \mathbf{e}_{-q} e^{i(-\boldsymbol{q}\cdot\boldsymbol{r}_n + \tilde{\omega}t_n)} \text{ and } \mathbf{e}_{m\pm} = \mathbf{e}_q e^{i(\boldsymbol{q}\cdot\boldsymbol{r}_m - \tilde{\omega}t_m)} \pm \mathbf{e}_{-q} e^{i(-\boldsymbol{q}\cdot\boldsymbol{r}_m - \tilde{\omega}t_m)}.$$

The cluster energy (3.8) at site n can then be given by

$$\mathcal{H}_n = -2\sigma_o^2 \sum_m J_{mn} \mathbf{e}_q \cdot \mathbf{e}_{-q} e^{i\{\boldsymbol{q}(\boldsymbol{r}_n - \boldsymbol{r}_m) - \tilde{\omega}(t_n - t_m)\}}.$$

The observable energy in timescale t_o can therefore be given by

$$\langle \mathcal{H}_n \rangle_{t_o} = \frac{1}{2t_o}\int_{-t_o}^{+t_o} \mathcal{H}_n \mathrm{d}(t_n - t_m) = \sigma_o^2 \Gamma \mathbf{e}_q \cdot \mathbf{e}_{-q} \sum_m J_{nm}(\boldsymbol{q})\, e^{i\boldsymbol{q}\cdot(\boldsymbol{r}_n - \boldsymbol{r}_m)}, \tag{3.9a}$$

where

$$J(\boldsymbol{q}) = \sum_{n,m} J_{mn}(\boldsymbol{q}) e^{i\boldsymbol{q}\cdot(\boldsymbol{r}_n - \boldsymbol{r}_m)} \tag{3.9b}$$

and

$$\Gamma = \frac{1}{2t_o}\int_{-t_o}^{+t_o} e^{-i\tilde{\omega}(t_n - t_m)} \mathrm{d}(t_n - t_m) = \frac{\sin \tilde{\omega} t_o}{\tilde{\omega} t_o} \tag{3.9c}$$

is the time parameter that is close to 1, if $\tilde{\omega} t_o < 1$.

To minimize $\langle \mathcal{H}_n \rangle_{t_o}$, the value of $J(\boldsymbol{q})$ should be restricted by the condition $\nabla_r J(\boldsymbol{q}) = 0$, by which the wavevector $\boldsymbol{q}$ can be specified for a stable cluster in the new phase. In addition, the factor $\mathbf{e}_q \cdot \mathbf{e}_{-q}$ in (3.9*a*) should be specified for the type of transitions.

Considering the relation $\mathbf{e}_{-q}^* = \mathbf{e}_{+q}$ for the unit vector, we have

$$\mathbf{e}_q^* \cdot \mathbf{e}_{-q} = \mathbf{e}_{-q}^* \cdot \mathbf{e}_q = 2\mathbf{e}_{\pm q}^2 = 2.$$

On the other hand $\mathbf{e}_n^* \cdot \mathbf{e}_n = \mathbf{e}_q \cdot \mathbf{e}_{-q} + \frac{1}{2}(e^{2i\boldsymbol{q}\cdot\boldsymbol{r}_n} + e^{-2i\boldsymbol{q}\cdot\boldsymbol{r}_n})$, hence we have either

$$e^{4i\boldsymbol{q}\cdot\boldsymbol{r}_n} = 1 \quad \text{or} \quad \mathbf{e}_q^2 = \mathbf{e}_{-q}^2 = 0. \tag{3.10}$$

The first case implies that $\boldsymbol{q} = 0$ and $\pm \boldsymbol{G}/2$ at $r_n = \pm \pi/G$, leading to another lattice arrangement of order variables over more than one unit cell. For example, figure 3.3

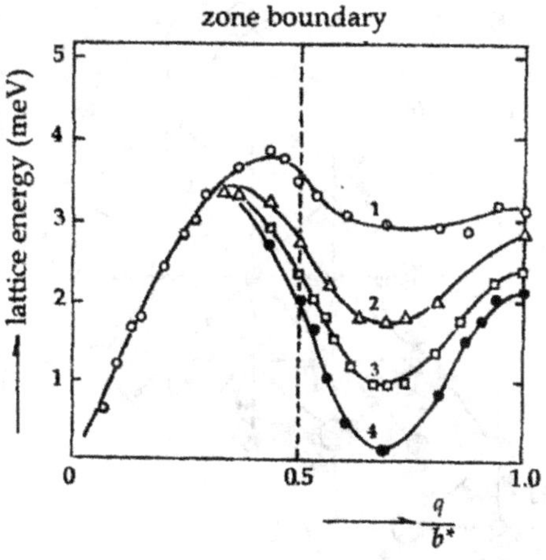

Figure 3.3. Phonon dispersion curve in K_2SeO_4 crystals near the Brillouin zone boundary obtained by neutron scattering experiments. Curves 1, 2, 3 and 4 were observed at 250, 175, 145 and 130 K, respectively. Reprinted with permission from [4], Copyright (1977) by the American Physical Society.

shows an example of phonon dispersion over the Brillouin zone in K_2SeO_4 crystals. The second relation indicates an *incommensurate distribution* of order variables in the lattice. For example, if $e_{q_z} = 0$, we have the relation $e_{q_x}^2 + e_{q_y}^2 = 1$, i.e. $e_{q_x} + ie_{q_y} = e^{i\theta}$, $0 \leqslant \theta \leqslant 2\pi$, giving an incommensurate distribution in crystals.

In the following, we discuss some examples of clusters to show how the direction of propagation can be determined by the lattice modulation.

Example 1: Cubic-to-tetragonal change

Figure 3.4 shows the schematic view of short-range correlations in a representative perovskite $SrTiO_3$ crystal, where octahedral TiO_3^{2+} ions are correlated in each other's nearest- and next nearest-neighbor distances, at unknown strengths J and J', respectively.

Regarding this cluster, (3.9*b*) can be written explicitly as

$$J(\boldsymbol{q}) = 2J(\cos q_a a + \cos q_b b + \cos q_c c) \\ + 4J'(\cos q_b b \cos q_c c + \cos q_c c \cos q_a a + \cos q_a a \cos q_b b).$$

Dealing with cubic-tetragonal transitions, it is convenient to assume orthorhombic a, b and c as above, and differentiate to set to zero for minimum, i.e.

$$\begin{aligned} &(\sin q_a a)(J + 2J' \cos q_b b + 2J' \cos q_c c) = 0, \\ &(\sin q_b b)(J + 2J' \cos q_c c + 2J' \cos q_a a) = 0 \\ \text{and} \quad &(\sin q_c c)(J + 2J' \cos q_a a + 2J' \cos q_b b) = 0. \end{aligned} \tag{3.11a}$$

From (3.11*a*), we can find the wavevector $\boldsymbol{q}_1$ to satisfy (3.11*a*) as determined by

$$\sin q_{1_a} a = \sin q_{1_b} b = \sin q_{1_c} c = 0 \quad \text{or} \quad \boldsymbol{q}_1 = \left(\frac{\pi}{a} l, \frac{\pi}{b} m, \frac{\pi}{c} n\right) = 0 \tag{3.11b}$$

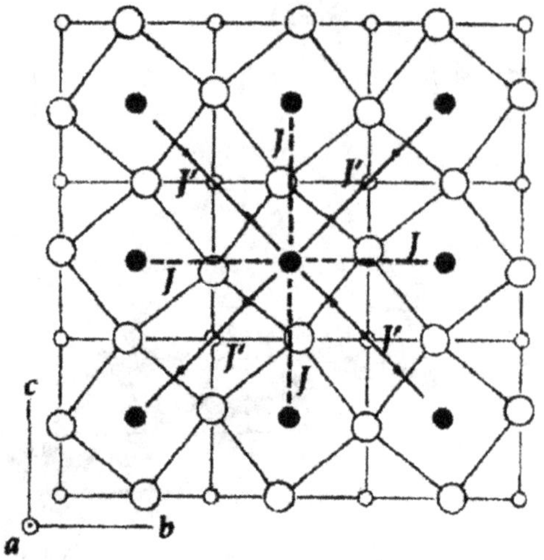

Figure 3.4. A model of the short-range cluster proposed for perovskites. Interaction parameters J and J' assigned to six nearest neighbors and 12 next-nearest neighbors, respectively. Reprinted by permission from Springer Nature [5] © 2010.

where l, m and n are zero or integers, indicating that $J(\boldsymbol{q}_1) = 6J + 12J'$ for 6 nearest- and 12 next-nearest neighbors, respectively.

On the other hand, from (3.11a) we obtain the relations

$$\cos q_{2_b}b + \cos q_{2_c}c = \cos q_{2_c}c + \cos q_{2_a}a = \cos q_{2_a}a + \cos q_{2_b}b = -1 + \frac{J}{2J'} \tag{3.12a}$$

indicating for the lattice arrangement of $\boldsymbol{q}_2$ is incommensurate along the direction, if $\left|-1 + \frac{J}{2J'}\right| \leqslant 2$.

In contrast, if choosing $\boldsymbol{q}_3$ from another set of relations

$$\sin q_{3_a}a = 0, \quad \cos q_{3_c}c + \cos q_{3_a}a = -1 + \frac{J}{2J'} \quad \text{and} \quad \cos q_{3_a}a + \cos q_{3_b}b = -1 + \frac{J}{2J'}, \tag{3.13}$$

the wavevector $\boldsymbol{q}_3 = \left(\frac{2\pi}{a}, q_{3b}, q_{3c}\right)$ is a two-dimensional incommensurate vector in the bc plane, where $\cos q_{3_b}b = \cos q_{3_c}c = -2 + \frac{J}{2J'}$ and $\left|-2 + \frac{J}{2J'}\right| \leqslant 1$. Experimentally, (3.13) is for orthorhombic to tetragonal transitions in perovskite crystals.

Example 2: Structural change in monoclinic TSCC

Figure 3.5 shows the molecular arrangement in monoclinic TSCC (tris-sarcosine calcium chloride) crystals in the hexagonal bc plane. A plausible cluster of pseudospins

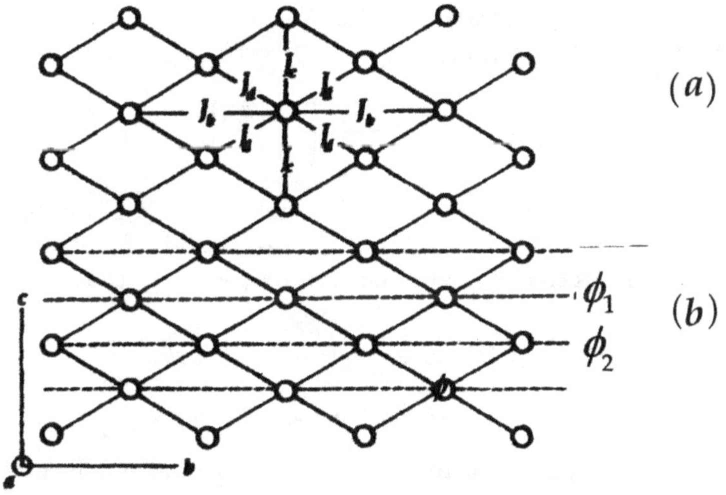

Figure 3.5. Pseudospin lattice in a TSCC crystal. (*a*) A cluster in *bc* plane. (*b*) Parallel adjacent chains of pseudospins along the *b* directions, intersecting in relation to a phase difference $\phi_1 - \phi_2$, where $0 \leqslant \phi \leqslant 2\pi$. Reprinted by permission from Springer Nature [5] © 2010.

consists of a Ca^{2+} ion surrounded in the octahedral form by six *sarcosine* molecules. The figure illustrates a model of such a cluster network in TSCC, where correlation parameters J_b, J_c and J_d are assigned to nearest- and next nearest-neighbors, respectively.

With these notations, the correlation function $J(\boldsymbol{q})$ is expressed as

$$J(\boldsymbol{q}) = 2J_a \cos q_a a + 2J_b \cos q_b b + 2J_c \cos q_c c + 4J_d \cos \frac{q_b b}{2} \cos \frac{q_c c}{2}, \qquad (3.14a)$$

from which we can derive specific wave vectors $\boldsymbol{q}_1$, $\boldsymbol{q}_2$, $\boldsymbol{q}_3$ to satisfy the relation $\nabla_q J(\boldsymbol{q}) = 0$; namely

$$q_1 = \left(\frac{\pi l}{a}, \frac{\pi m}{b}, \frac{\pi n}{c}\right) \text{ where } (l,\ m,\ n) \text{ are 0 or integers}$$

are commensurate with the lattice, and $J(\boldsymbol{q}_1) = 2J_a + 2J_b + 2J_c$; while

$$\boldsymbol{q}_2 = \left(\frac{\pi l}{a}, q_{2b}, \frac{\pi n}{c}\right) \quad \text{where} \quad \cos \frac{q_{2b} b}{2} = -\frac{J_d}{2J_b}$$

and

$$\boldsymbol{q}_3 = \left(\frac{\pi l}{a}, \frac{\pi m}{b}, q_{3c}\right) \quad \text{where} \quad \cos \frac{q_{3c} c}{2} = -\frac{J_d}{2J_c}.$$

With these $\boldsymbol{q}_2$ and $\boldsymbol{q}_3$, waves are incommensurate along b and c directions, if

$$\left|\frac{J_d}{2J_b}\right| \leqslant 1 \quad \text{and} \quad \left|\frac{J_d}{2J_c}\right| \leqslant 1.$$

A TSCC crystal at temperatures below 120 K is known as ferroelectric along the b direction, for which the wavevector $\boldsymbol{q}_2$ is responsible, and

$$J(\boldsymbol{q}_2) = 2J_a + 2J_c + 2J_b\left(1 - \frac{J_d^2}{J_b^2}\right). \tag{3.14b}$$

Based on the above argument, some useful remarks emerge on the correlation scheme. Namely, if $J_d = 0$, we have $J(\boldsymbol{q}_1) = J(\boldsymbol{q}_2)$; on the other hand, if $J_d = -2J_b$, we obtain $J(\boldsymbol{q}_1) > J(\boldsymbol{q}_2)$, thereby the incommensurate $\boldsymbol{q}_2$ gives a lower energy than the commensurate $\boldsymbol{q}_1$.

Furthermore, writing $\varphi = \frac{q_2 b}{2}$ as an angular variable, (3.14b) is expressed as

$$J(\boldsymbol{q}_2) = 2J_a + 2J_c + 2J_d \cos 2\varphi + 4J_d \cos \varphi, \tag{3.14c}$$

which is a formula familiar in the theory of magnetism for spin–spin arrangement in one-dimension. A similar situation occurs to charge-density waves, as the interchain correlation was assumed by Rice [2] as $J_d \propto \cos(\varphi_1 - \varphi_2)$.

3.3 Macroscopic views of structural phase transitions

3.3.1 Landau's mean field theory

We consider an order variable discussed in previous sections to represent thermodynamic variables. The thermodynamic average of variable $\boldsymbol{\sigma}$ expressed by $\langle\boldsymbol{\sigma}\rangle$ corresponds to order parameter η, i.e. $\langle\boldsymbol{\sigma}\rangle = \eta$.

Historically, a structural phase transition in crystals was first observed from specific heat anomalies measured under varying temperature. Under constant volume and pressure, the specific heat curve measured as a function of temperature T showed an anomaly like the letter Greek-lambda λ, as illustrated in figure 3.6; and normally called a λ-transition.

Landau considered Gibbs' free energy $G(\eta)$ to explain the λ-anomaly by the order parameter η by expanding $G(\eta)$ into a series of η, as shown as follows.

$$G(\eta) = G(0) + \frac{1}{2}A\eta^2 + \frac{1}{4}B\eta^4 + \cdots \tag{3.15a}$$

where the coefficients A, B, … are constant or functions of p and T. Under the condition for constant volume, however, p is kept constant. And we assume

$$G(\eta) = G(-\eta). \tag{3.15b}$$

In an equilibrium crystal with surroundings, the difference should be minimum against $\Delta\eta$ for equilibrium at T, we set

$$\Delta G = \left(\frac{\partial G}{\partial \eta}\right)_T \Delta\eta = \frac{1}{2}A\eta^2 + \frac{1}{4}B\eta^4 \geqslant 0$$

to take a minimum value against a change $\Delta\eta = \eta - \eta_o$, therefore

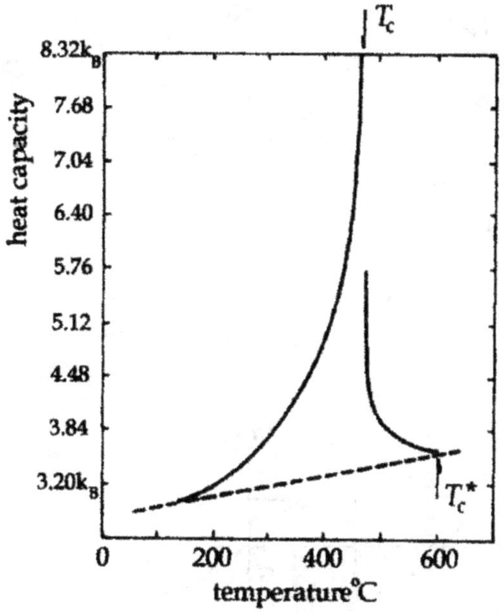

Figure 3.6. A transition anomaly in C_V observed in β-brass, signified by a sharp rise at T_c with narrow threshold T_c^*, and a gradual decay for $T < T_c$.

$$\eta_0 = 0 \quad \text{or} \quad \eta_0^2 = -\frac{A}{B}.$$

The first solution represents a disordered state, while the second is partially ordered if $A \neq 0$, otherwise disordered. If $A = 0$, both are the same. Hence, the factor $A \neq 0$ is temperature-dependent, for which Landau proposed the following relations:

$$\begin{aligned} A &= A'(T - T_0) \quad &\text{for} \quad &T > T_0 \\ A &= A'(T_0 - T) \quad &\text{for} \quad &T < T_0 \end{aligned}. \tag{3.16}$$

Considering both A' and B are positive in (3.16), the above η_0 is in parabolic form, and the Gibbs function for η_0 can be expressed by

$$G(\eta_o = 0) = G(0) \quad \text{for} \quad T > T_0 \tag{3.17a}$$

and

$$G(\eta_0) = G(0) - \frac{3A^2}{4B} \quad \text{and} \quad \eta_0 = \pm\sqrt{\frac{A'}{B}(T_0 - T)} \quad \text{for} \quad T < T_0. \tag{3.17b}$$

Figure 3.7 shows schematically such a change of heat capacity related with Gibbs' function with respect to temperature.

Figure 3.7(a) is for (3.13a), where $\Delta G(\eta)$ around η_0 is infinitesimal at all temperatures for $T > T_0$, whereas in figures 3.7(b) and 3.7(c) two minima at $\pm\eta_0$ are shifted in parabolic symmetry with decreasing temperature.

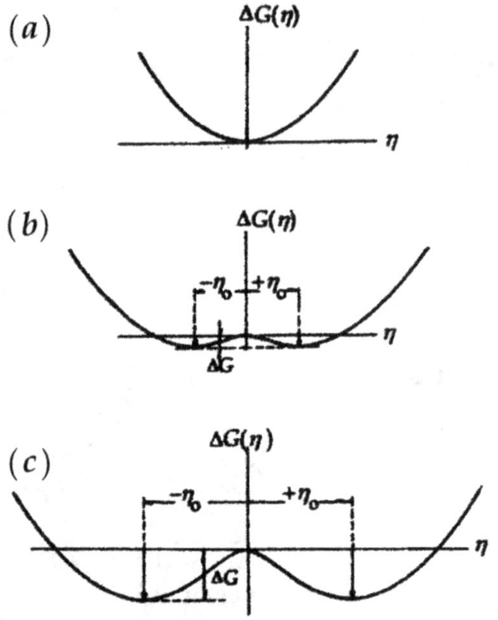

Figure 3.7. Binary fluctuations of Gibbs' potentials $\pm\,\eta_o$, (*a*) at T_c, (*b*) and (*c*) for $T < T_c$, showing increasing minimum depths on decreasing temperature.

3.3.2 The Curie–Weiss law of susceptibility

It is interesting to note that the concept of Weiss field is built in (3.11). To minimize for equilibrium, we expand Gibbs' function mathematically with respect to an arbitrary variation of η. Similarly, we analyze Landau's expansion for $T > T_o$.

Writing $\Delta G = \frac{A}{2}\eta^2 + \frac{B}{4}\eta^4 + \cdots = \langle \Delta U_n \rangle = -\eta X_{\text{int}}$, we obtain

$$X_{\text{int}} = A\eta + B\eta^3 + \cdots = -\left\langle \frac{\partial U(\sigma_n)}{\partial \sigma_n} \right\rangle,$$

which is identical to $X_{\text{int}}(\eta) = -\frac{\partial U}{\partial \eta}$ where η and U are in phase with equilibrium crystals. Accordingly, X_{int} is an internal field in mean-field accuracy, acting on η as if applied externally, being essential for Weiss' molecular field in ferromagnetic crystals (1907).

We evaluate the relation $\Delta G = 0$ for equilibrium at temperatures above T_o, assuming small variation $\delta\eta$ in the close vicinity, and derive the Curie–Weiss law of susceptibility.

Writing

$$\Delta G_{\,T>T_o} = \Delta G(\eta) - \eta X_{\text{int}} \quad \text{and} \quad \left(\frac{\partial \Delta G_{\,T>T_o}}{\partial \eta}\right)_T = 0,$$

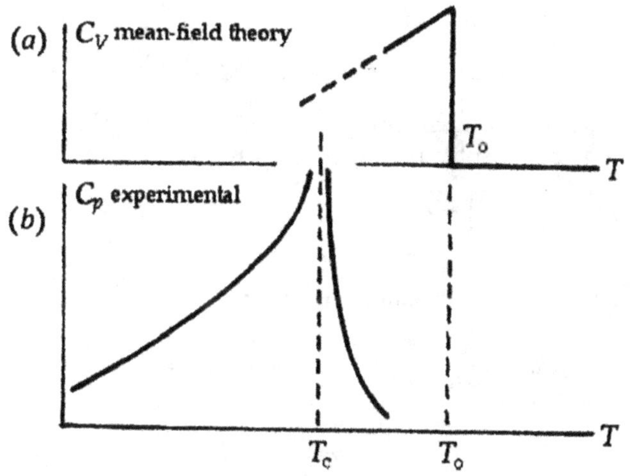

Figure 3.8. A comparison of typical specific heat anomalies, where (*a*) mean-field theory and (*b*) experimental anomalies. Reprinted by permission from Springer Nature [5] © 2010.

we drive $A\eta - X_{\text{int}} = 0$, and the susceptibility formula can be obtained as

$$\chi_{T>T_o} = \frac{\eta}{X_{\text{int}}} = \frac{1}{A} = \frac{1}{A'(T - T_o)}. \tag{3.17}$$

Similarly at $T = T_o$, we obtain $\chi_{T_o} = 0$ by assuming $\eta_o = 0$.

Here, the transition temperature is designated to T_o in Landau's mean-field theory, however corrected substantially by the observed transition anomaly. The second-order phase transition is signified by a discontinuous change of $\frac{1}{2}\left(\frac{\partial^2 G}{\partial \eta^2}\right)_{T_o}$ at $T = T_o$, as illustrated in figure 3.8.

3.3.3 Critical fluctuations

Experimentally, a second-order phase transition is observed at the *critical temperature* T_c determined by the specific heat anomaly. It is signified by the clustered pseudospins $\boldsymbol{\sigma}_q = \boldsymbol{\sigma}_o f(\phi)$ pinned bythenegativepotential $-\Delta U_q(\phi) \approx -\frac{a}{2}\sigma(\phi)^2$ at T_c. Resulting from the phasing process, it is in a sufficiently large magnitude for transition, and both are functions of phase ϕ in the system moving at a speed $v = \frac{\bar{\omega}}{q}$ between ϕ_+ and ϕ_-; $\phi = 0$ determines the initial generation of nonlinearity.

Assuming fluctuations between $\Delta U_q(\phi_+)$ and $\Delta U_q(\phi_-)$, as shown in figure 3.7, we consider the adiabatic potential $\Delta U_q(\phi)$ fluctuating in the vicinity of $\phi = 0$. At the crossing point, these two potentials $\frac{a}{2}\sigma_o^2(\phi)$ are degenerate in this approximation, but lifted by a symmetric perturbation expressed by $\Delta U_q(\phi) = Vf(\phi)$, where V is a constant and $f(\phi)$ a normalized even function for $-\frac{\pi}{2} \leqslant \phi \leqslant +\frac{\pi}{2}$. Such a problem of mesoscopic fluctuation can be discussed in analogy with level-crossing in quantum mechanics.

First, we consider a linear combination

$$\sigma = c_+\sigma_+ + c_-\sigma_- \quad \text{where } c_+^2 + c_-^2 = 1.$$

Here, c_1 and c_2 are the normalization factor for $\sigma(\pm\Delta\phi)$. With the perturbation of $Vf(\phi)$ to the energy ε, we solve the secular equation

$$\begin{vmatrix} +\varepsilon_+ - \varepsilon & \Delta \\ \Delta & -\varepsilon_- - \varepsilon \end{vmatrix} = 0,$$

from which eigenvalue ε can be calculated with

$$\varepsilon_+, \ \varepsilon_- = \frac{1}{\pi}\int_{-\pi/2}^{+\pi/2} \sigma_o^* \cdot Vf(\phi)\,\sigma_o\, \mathrm{d}\phi = V\sigma_o^2 = \varepsilon_o$$

and

$$\Delta = \frac{1}{\pi}\int_{-\pi/2}^{+\pi/2} \sigma_+^* \cdot Vf(\phi)\,\sigma_- \mathrm{d}\phi = \frac{V\sigma_o^2}{\pi}\int_{-\pi/2}^{+\pi/2} f(\phi)\cos 2\phi\, \mathrm{d}\phi = \frac{V\sigma_o^2}{2},$$

and obtain

$$\varepsilon = \varepsilon_o \pm \Delta. \tag{3.18}$$

Accordingly, the degeneracy is lifted by the energy gap 2Δ, as shown in figure 3.9, corresponding to $\sigma_q \rightleftarrows -\sigma_{-q}$. Therefore, the eigenvalues ε_A and ε_P for symmetric and antisymmetric fluctuations are expressed as

$$\sigma_A = \frac{\sigma_+ + \sigma_-}{\sqrt{2}} \quad \text{and} \quad \sigma_P = \frac{\sigma_+ - \sigma_-}{\sqrt{2}}, \tag{3.19}$$

which are called *amplitude- and phase-modes*, respectively, representing symmetric and antisymmetric fluctuations. The gap 2Δ between σ_A and σ_P signifies thepotential $V(0)$ that emerges at T_c, to initiate nonlinearity on lowering temperature.

Writing $\sigma_A = \sigma_o \cos\phi$ and $\sigma_P = \sigma_o \sin\phi$, (3.16) indicates that σ_A and σ_P are components of a classical vector $\sigma = (\sigma_A, \ \sigma_P)$. The critical fluctuations can therefore be discussed in a complex form

$$\mp i\sigma_A + \sigma_P = \sigma_o e^{i\left(\phi \pm \frac{\pi}{2}\right)}, \tag{3.20}$$

from which we know σ_P, σ_P are determined at $\phi = 0, \ \frac{\pi}{2}$, respectively.

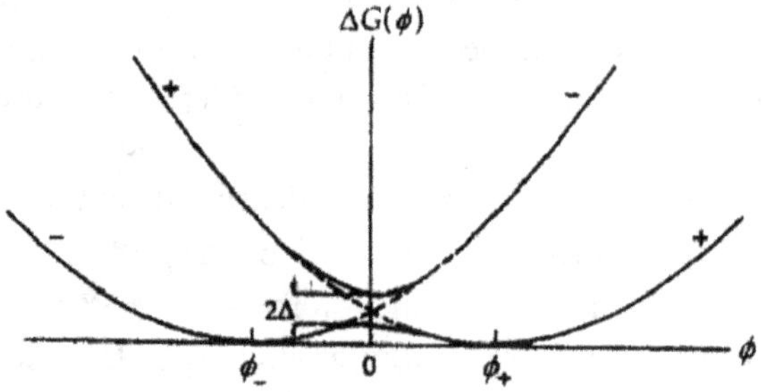

Figure 3.9. Phase fluctuations in Gibbs' potential near the critical temperature T_c.

The order variable $\boldsymbol{\sigma}$ defined as a classical vector is well documented by experimental work including neutron inelastic and paramagnetic resonance [1, 2]. Therefore, its existence in mesoscopic nature is no longer hypothetical. We therefore proceed to discuss thermodynamics of $\boldsymbol{\sigma}$ for entropy production in the next section.

3.3.4 Entropy production at critical temperatures

Now that local displacements of constituents are considered for order in given crystals, which are responsible for structural transformation. We should discuss this mechanism for entropy production associated with heat transfer between the two systems. As discussed in section 3.1, the crystal lattice represented by phonon vibration should be considered primarily as a heat reservoir of excess energy originating from the system of order variables.

Noticing that the lattice is dynamically conservative, the energy transfer can be dealt with for matrix elements between different soliton states described by *phonon wavefunctions* $\langle \boldsymbol{K};\ \boldsymbol{q}, \tilde{\omega}|$ and $|\boldsymbol{K}';\ \boldsymbol{q}', \tilde{\omega}'\rangle$, and the off-diagonal matrix element

$$\langle \boldsymbol{K}; +\boldsymbol{q}, \tilde{\omega}|\sigma_{\mathrm{A}}|\boldsymbol{K}'; -\boldsymbol{q}', \tilde{\omega}'\rangle \neq 0 \qquad \text{for} \quad \boldsymbol{q} + \boldsymbol{q}' \neq 0$$

should be dealt with for transitions at a given temperature.

With the nonzero element (3.18), the phonon scattering can be specified as inelastic scattering between phonon states, for which the momentum conservation relation can be written as

$$\boldsymbol{K} \pm \boldsymbol{q} = \boldsymbol{K}' \mp \boldsymbol{q}' \quad \text{or} \quad \boldsymbol{K}' - \boldsymbol{K} = \pm\, 2\boldsymbol{q}, \tag{3.21}$$

for which the corresponding energy relation is indicated as inelastic $\Delta\varepsilon_{\pm q} \neq 0$.

The energy transfer processes can be described by using non-diagonal elements of perturbation from phonon-soliton interaction. Expressing this as $\mathcal{H}'(\boldsymbol{K},\ \boldsymbol{K}') = \lambda\sigma_{q,-q}$, the equation for phonon scattering can be written as

$$\mathcal{H}_{-q}(\boldsymbol{K}') = \mathcal{H}_q(\boldsymbol{K}) + \lambda\sigma_q, \tag{3.22}$$

where $\mathcal{H}_q(\boldsymbol{K})$ and $\mathcal{H}_{-q}(\boldsymbol{K}')$ are unperturbed and perturbed Hamiltonians, respectively.

Starting with Schrödinger's equation

$$-\hbar^2\ddot{\psi}_q(\boldsymbol{K}) = \mathcal{H}_q(\boldsymbol{K})\psi_q(\boldsymbol{K}) \quad \text{and} \quad -\hbar^2\ddot{\psi}_{-q}(\boldsymbol{K}') = \mathcal{H}_{-q}(\boldsymbol{K}')\psi_{-q}(\boldsymbol{K}')$$

written for $\psi_q(\boldsymbol{K}) = a_q(t)u_q(\boldsymbol{K})e^{-i\tilde{\omega}t}$, equation (3.19) can be solved for the perturbed $\psi_{-q}(\boldsymbol{K}')$ from

$$i\hbar\dot{a}_{-q}e^{-i\tilde{\omega}_{-q}t} = a_q e^{-i\tilde{\omega}_{-q}t}\frac{\lambda}{2t_{\mathrm{o}}}\int_{-t_{\mathrm{o}}}^{+t_{\mathrm{o}}} u_{-q}^*(\boldsymbol{K}')\sigma_{\mathrm{A}}u_q(\boldsymbol{K})\,\mathrm{d}t,$$

and we have

$$a_{-q} = \frac{\lambda}{i\hbar}\langle -\boldsymbol{q}|\sigma_{\mathrm{A}}|\boldsymbol{q}\rangle a_q e^{i(\tilde{\omega}_{-q} - \tilde{\omega}_q)t}$$

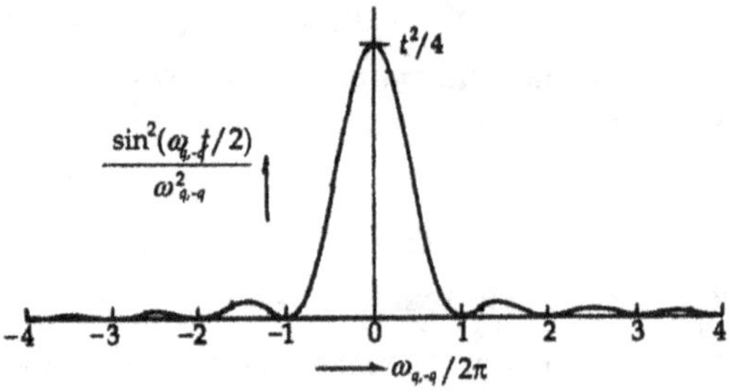

Figure 3.10. A relative intensity spectrum of energy transfer to the lattice, determined by (3.24).

where

$$\langle -\boldsymbol{q}|\sigma_{\mathrm{A}}|\boldsymbol{q}\rangle = \frac{1}{2t_o}\int_{-t_o}^{+t_o} u_{-q}^*(\boldsymbol{K}')\langle -\boldsymbol{q}|\sigma_{\mathrm{A}}|\,\boldsymbol{q}\rangle u_q(\boldsymbol{K})\,\mathrm{d}t, \tag{3.23a}$$

and

$$\frac{1}{2t_o}\int_{-t_o}^{+t_o} u_{-q}^*(\boldsymbol{K}')u_{+q}(\boldsymbol{K})\,\mathrm{d}t = 1 \tag{3.23b}$$

which is noted as modified by lattice phonon distributions at the critical condition, providing a substantial magnitude.

The probability for inversion should be calculated from $\lambda\sigma_q$ in (3.19), considering $t = t_o$ for thermodynamic phase change in practice. Hence, the expression

$$a_{-q}^*(t)a_q(t) = \frac{4\langle -\boldsymbol{q}|\sigma_q|\boldsymbol{q}\rangle^2 \sin^2\frac{(\tilde{\omega}_{-q} - \tilde{\omega}_q)t}{2}}{\hbar^2(\tilde{\omega}_{-q} - \tilde{\omega}_q)^2} \tag{3.24}$$

characterizes the probability for inversion $\boldsymbol{q} \rightleftarrows -\boldsymbol{q}$ associated with phase transitions, as shown in figure 3.10.

It is noted however that (3.21) represents the probability for σ_{A} measured in timescale of phonons, whose order of magnitude is comparable solitons in the mesoscopic timescale. However, we can assume that the phonon number at T_c is much larger than the soliton number in the thermodynamic region of phase transitions. Therefore, while proportional to $\langle -\boldsymbol{q}|\sigma_{\mathrm{A}}|\boldsymbol{q}\rangle^2$, the probability is dominated by phonon number n_K and $n_{K'}$ at T_c for phase transitions originating from clustered seeds. Hence, the gap Δ between σ_{A} and σ_{P} in molar units is determined by the relation

$$\mathrm{k_B}T_c = 2\,\Delta, \tag{3.25}$$

specified by a difference between soliton numbers $|n_q - n_{-q}|$.

3.4 Observing critical anomalies

3.4.1 Amplitude anomalies

Spectroscopically, critical anomalies are detected clearly from diffuse patterns of x-ray diffractions, however the resolution is not adequate. In contrast, inelastic

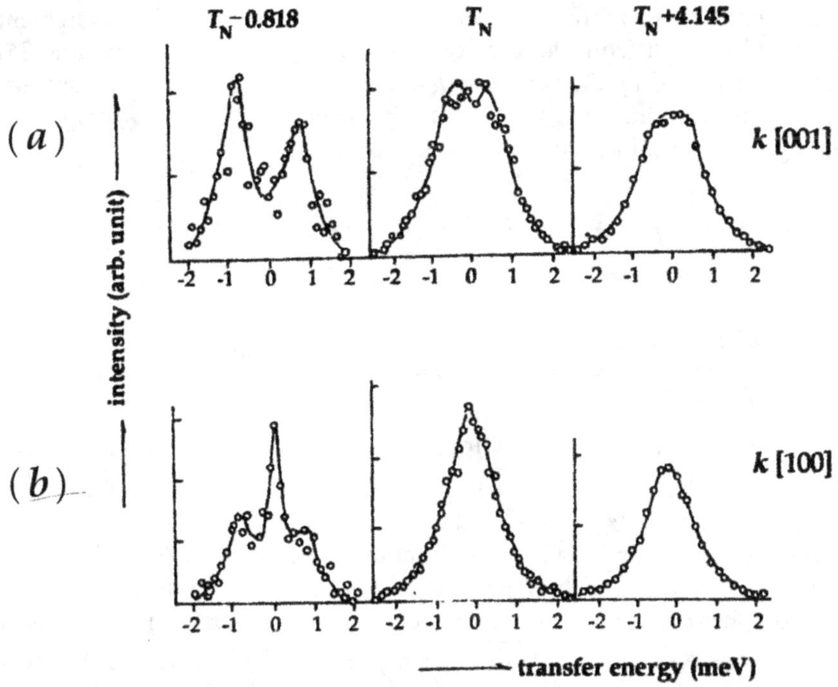

Figure 3.11. Inelastically scattered neutrons from magnetic MnF_2 crystals detected at Neél temperature T_N, where A- and P-peaks are recognized in different proportion, depending on the scattering geometry with respect to crystal axes. Reproduced with permission from [1]. Copyright (1970) by the American Physical Society.

scattering diffraction of thermal neutrons from magnetic crystals, such as MnF_2, exhibits anomalous scattering intensities, as shown in figure 3.11 [1].

In neutron inelastic experiments on modulated crystals, the structural form factors of magnetic spins are modulated as

$$|f(\boldsymbol{G}_z \pm \boldsymbol{q})|^2 = |f(\boldsymbol{G}_z)|^2 \pm 2i\boldsymbol{q} \cdot \nabla_q |f(\boldsymbol{G}_z)|^2$$

where $\boldsymbol{G}_z$ is a reciprocal vector at the critical point near $\pm\frac{G}{2}$, and the anomalous intensity shows an intensity distribution

$$\langle \Delta I(\boldsymbol{G}_z \pm \boldsymbol{q}_z) \rangle = \mathrm{A}\langle \cos(\boldsymbol{q}_z \cdot \boldsymbol{r}) \rangle + \mathrm{P}\langle \sin(\boldsymbol{q}_z \cdot \boldsymbol{r}) \rangle$$

where $\mathrm{A} \propto \int_{v(r)} |f(\boldsymbol{G}_z)|^2 \cos(\boldsymbol{G}_z \cdot \boldsymbol{r})\, \mathrm{d}^3\boldsymbol{r}$ and $\mathrm{P} \propto \int_{V(r)} 2\boldsymbol{q} \cdot \nabla_q |f(\boldsymbol{G}_z)|^2 \cos(\boldsymbol{G}_z \cdot \boldsymbol{r})\, \mathrm{d}^3\boldsymbol{r}$, representing fluctuations of symmetric and antisymmetric intensities, respectively.

It is noted that there are clear evidence for these A and P modes in the figure, depending on orientations of the sample crystal.

3.4.2 Frequency scanning of phase anomalies

Critical anomalies described by (3.16) are expressed by σ_A and σ_P in terms of their *phase* ϕ. Experimentally, these can be studied by scanning the variation in space and time. In practice, it is important to consider the timescale t_o of measurements [3], which should be sufficiently long as compared with a characteristic time $2\pi/\tilde{\omega}$ of fluctuations, specified by the condition $\tilde{\omega}t_o > 1$ to interpret fluctuation spectra.

Considering sampling experiments of timescale t_o, the time average of these fluctuating order variables can be expressed by

$$\langle\sigma_A\rangle_t = \frac{\sqrt{2}\sigma_o}{2t_o}\int_{-t_o}^{+t_o}\cos(\boldsymbol{q}\cdot\boldsymbol{r} - \tilde{\omega}t)\,dt = 2\sqrt{2}\sigma_o\frac{\sin\tilde{\omega}t_o}{\tilde{\omega}t_o}\cos(\boldsymbol{q}\cdot\boldsymbol{r}) \qquad (3.26a)$$

and

$$\langle\sigma_P\rangle = \frac{\sqrt{2}\sigma_o}{2t_o}\int_{-t_o}^{+t_o}\sin(\boldsymbol{q}\cdot\boldsymbol{r} - \tilde{\omega}t)\,dt = 2\sqrt{2}\sigma_o\frac{\sin\tilde{\omega}t_o}{\tilde{\omega}t_o}\sin(\boldsymbol{q}\cdot\boldsymbol{r}), \qquad (3.26b)$$

for which we define $\Gamma = \frac{\sin\tilde{\omega}t_o}{\tilde{\omega}t_o}$ for the reducing factor from $2\sqrt{2}\sigma_o$, and for $\tilde{\omega}t_o \to 1$ we have $\Gamma \to 1$. Accordingly, these time averages can be examined for the spatial phase $\boldsymbol{q}\cdot\boldsymbol{r}$.

Defining $\langle\sigma_A\rangle_t = \sigma_A$ and $\langle\sigma_P\rangle_t = \sigma_P$, the complex order variable (3.17) can be used for experimental analysis. Observed spectra of σ_A and σ_P are usually scanned as a function of phase ϕ, and conveniently deal with another variable ξ, using the relations $\xi = \cos\phi$, hence $\sqrt{1-\xi^2} = \sin\phi$. Figure 3.12 illustrates the spectra as a function of ξ, and figure 3.13 shows schematic anomalies in specific heat, which is compared with the mean-field theory.

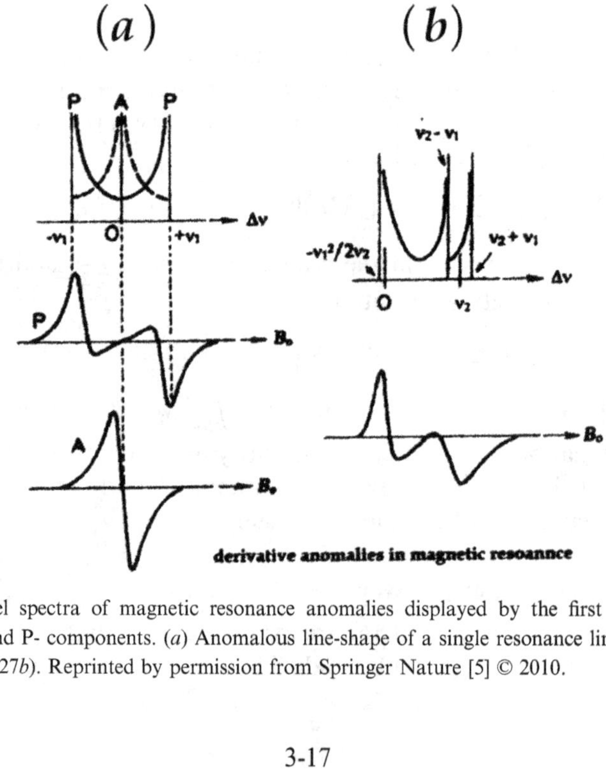

Figure 3.12. Model spectra of magnetic resonance anomalies displayed by the first derivative $d\chi''/dB_o$, composed of A- and P- components. (*a*) Anomalous line-shape of a single resonance line, and (*b*) from two combined lines, (3.27*b*). Reprinted by permission from Springer Nature [5] © 2010.

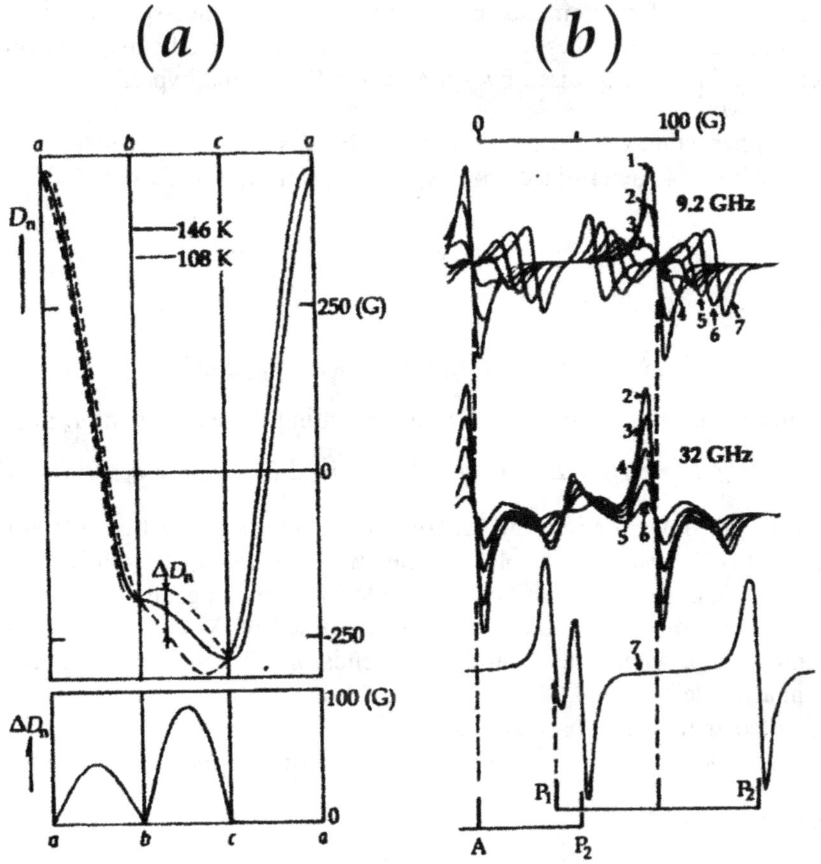

Figure 3.13. (*a*) An angular dependence of the fine structure D_n of Mn^{2+} spectra in TSCC. (*b*) Critical anomalies in 55Mn–hyperfine lines. Here, the splitting leading to domain-lines exhibit anomalous shape, becoming normal splitting indicated by 1, 2, 3, …, 7.

Applying paramagnetic resonance technique to study binary transitions due to the complex variable, we use a small amount of impurity ions of iron- or rare-earth groups as test particles in crystals for obtaining magnetic resonance spectra. In this technique, the impurity spin $\boldsymbol{S}$ is considered to be modified by changes in surrounding lattice in the form, $\boldsymbol{S}' = \boldsymbol{\alpha}\,\boldsymbol{S}$, where $\boldsymbol{\alpha} = \mathbf{1} + \sigma\mathbf{e}$ composed by identity- and local strain-tensors. In fact, the ionic magnetic moment originates from electronic orbital angular momentum $\mathbf{L}$, which is quenched by the surrounding crystal field. Therefore, the effective magnetic moment can be expressed with the Landé factor as $-\beta g\boldsymbol{S}$ where $\beta = 9.27 \times 10^{-24}$ joule/tesla.

The modified spin-Hamiltonian can then be expressed as $\mathcal{H}' = \mathcal{H} + \mathcal{H}_1$ where

$$\begin{aligned} \mathcal{H}' &= -\beta\langle \boldsymbol{S}'|\boldsymbol{g}|\boldsymbol{B}_o\rangle + \langle \boldsymbol{S}'|\boldsymbol{D}|\boldsymbol{S}'\rangle + \langle \boldsymbol{S}'|\boldsymbol{K}|\boldsymbol{I}\rangle, \\ \mathcal{H} &= -\beta\langle \boldsymbol{S}|\boldsymbol{g}|\boldsymbol{B}_o\rangle + \langle \boldsymbol{S}|\boldsymbol{D}|\boldsymbol{S}\rangle + \langle \boldsymbol{S}|\boldsymbol{K}|\boldsymbol{I}\rangle \quad \text{and} \qquad (3.27a) \\ \mathcal{H}_1 &= -\sigma\,\beta\langle \boldsymbol{S}|\bar{\mathbf{e}}\boldsymbol{g}|\boldsymbol{B}_o\rangle + \sigma\langle \boldsymbol{S}|\bar{\mathbf{e}}\boldsymbol{D} + \boldsymbol{D}\mathbf{e}|\boldsymbol{S}\rangle + \sigma^2\langle \boldsymbol{S}|\bar{\mathbf{e}}\boldsymbol{D}\mathbf{e}|\boldsymbol{S}\rangle + \sigma\langle \boldsymbol{S}|\boldsymbol{K}|\boldsymbol{I}\rangle, \end{aligned}$$

which are essential for anomalies in paramagnetic resonance spectra from modulated crystals. In practice, these Zeeman terms can be specified by magnetic quantum number M and m, as expressed by $-g\beta MB_o$ for $\boldsymbol{S}$, and the hyperfine interaction is written as KMm.

On the other hand, the direction $\boldsymbol{n}$ nuclear spin $\boldsymbol{I}$ is parallel to $\boldsymbol{S}$, so that so-called hyperfine splitting is determined from $K_n^2 = \langle \boldsymbol{n}|\boldsymbol{\alpha}^\dagger \boldsymbol{K}^\dagger \boldsymbol{K}\boldsymbol{\alpha}|\boldsymbol{n}\rangle$ and expressed by

$$\Delta K_n' = \frac{2\sigma}{K_n}\langle \boldsymbol{n}|\mathbf{e}^\dagger \boldsymbol{K}^{\dagger 2} + \boldsymbol{K}^2\mathbf{e}|\boldsymbol{n}\rangle.$$

Further, from (3.23), we have the fine structure anomaly

$$\mathcal{H}_{1F} = \sigma\langle S_n|\mathbf{e}^\dagger \boldsymbol{D}^\dagger + \boldsymbol{D}\mathbf{e}|S_n\rangle + \sigma^2\langle S_n|\mathbf{e}^\dagger \boldsymbol{D}\mathbf{e}|S_n\rangle,$$

leading to the magnetic resonance formula by setting these factors as a_n and b_n, i.e.

$$\hbar\omega = g_n\beta B + (D_n \pm \Delta D_n)(2M+1) \quad \text{where} \quad \Delta D_n = a_n\sigma + b_n\sigma^2. \tag{3.27b}$$

Figure 3.13 [6] shows the intensity distribution of the fine structure ΔD_n for $\boldsymbol{B}_o \| \boldsymbol{n}$. Figure 3.13(*a*) [6] displays the plot of fine structure obtained from Mn^{2+}–doped crystals of TSCC (tris-sarcosine calcium chloride), showing anisoscopic splitting. Figure 3.13(*b*) shows transition anomalies composed of A- and P-modes, varying with temperature, where the separation depends on observing frequencies, and is clearly analyzable.

It is noted that these observed transition anomalies provide clear evidence for modulating order variables, substantiating as real objects, not a hypothesis.

Exercises

1. Discuss the amplitude mode σ_A. Why can this mode of order variable be responsible for entropy production processes?
2. The inversion $\boldsymbol{q} \rightleftarrows -\boldsymbol{q}$ in this chapter is mesoscopically applicable to infinite crystals. On the other hand, specified by surfaces, finite crystals cannot be homogeneous inside. In this chapter, such an inversion is considered as related to symmetry change, but should there be another type of singularity inside inhomogeneous crystals? Discuss this conflicting issue on inversion, if applied to practical crystals, while it is an acceptable model within limited range.

References

[1] Haken H 1973 *Quantenfeldtheorie des Festkorpers* (Stuttgart: B.G. Teubner) ch 6

[2] Schulhof M P, Heller P, Nathans R and Linz A 1970 *Phys. Rev.* B **1** 2403

[3] Rice M J 1978 *Charge Density Waves in Solitons and Condensed Matter Physics* ed A R Bishop and T Schneider (Berlin: Springer)

[4] Fujimoto M, Jerzak S and Windsch W 1986 *Phys. Rev.* B **34** 1668

[5] Iizumi M, Axe J D, Shirane G and Shimaoka K 1977 *Phys Rev.* B **15** 4392

[6] Fujimoto M 2010 *Thermodynamics of Crystalline States* (Berlin: Springer)

IOP Publishing

Introduction to the Mathematical Physics of Nonlinear Waves (Second Edition)

Minoru Fujimoto

Chapter 4

Soft modes of lattice displacements

Exhibiting structural changes, *soft modes* detected in dielectric experiments are characterized by *temperature-dependent frequencies* and *thermal relaxation* in crystals. Such experiments were performed with time-dependent electric field applied at the Brillouin-zone center, while in cell-doubling perovskites, soft modes were studied with neutron inelastic scatterings at zone boundaries. Observed as a function of temperature under constant volume, frequencies were temperature dependent. Showing an isothermal profile of collective displacements, soft modes represent the response from finite displacements. Reviewing reported data in this chapter, we discuss the symmetry change of order variables during structural transitions.

4.1 The Lyddane–Sachs–Teller relation

In ionic crystals, the polarization $\boldsymbol{P}(\boldsymbol{r}, t)$ is due to charge displacements in the lattice. Detected by an applied electric field $\boldsymbol{E} = \boldsymbol{E}_0 e^{-i\omega t}$ where $\boldsymbol{E}_0$ and ω are the amplitude and frequency, respectively, and the polarization in crystals occurs not only as $\alpha\boldsymbol{E}$, where α is the ionic polarizability, but also related to its displacement $\boldsymbol{u}(\boldsymbol{r}, \boldsymbol{t})$. It should be realized that these intrinsic variables $\boldsymbol{P}(\boldsymbol{r}, \boldsymbol{t})$ and $\boldsymbol{u}(\boldsymbol{r}, t)$ represent collective motion of microscopic variables as continuous functions in space–time.

Following Elliot and Gibson [1], we write the relation

$$\boldsymbol{P} = a\boldsymbol{u} + b\boldsymbol{E}, \tag{4.1}$$

which was originally proposed by Born and Huang [2], and the equation

$$\frac{\partial^2 \boldsymbol{u}}{\partial t^2} = a'\boldsymbol{u} + b'\boldsymbol{E} \tag{4.2}$$

doi:10.1088/978-0-7503-3759-5ch4

describes the lattice movement in sufficiently weak $\boldsymbol{E}$, where a, b and a', b' are constant coefficients for all dielectric crystals to be *electro-elastic*, as crystals are strained by polarization.

The vector $\boldsymbol{u}(\boldsymbol{r}, t)$ is either *irrotational* or *rotational* as signified by curl $\boldsymbol{u} \neq 0$ or curl $\boldsymbol{u} = 0$, respectively. The former applies to a case of laminar flow characterized by div $\boldsymbol{u} \neq 0$. Moreover, $\boldsymbol{P}$ and $\boldsymbol{u}$ are proportional to $e^{i(\boldsymbol{q}\cdot\boldsymbol{r}-\tilde{\omega}t)}$ for $\boldsymbol{E} = \boldsymbol{E}_o e^{i\omega t}$, hence $\boldsymbol{P} = \boldsymbol{P}_o e^{i(\boldsymbol{q}\cdot\boldsymbol{r}-\omega t)}$. In this case, writing the amplitude of $\boldsymbol{u}$ as $\boldsymbol{u}_o$, we have the relation

$$\boldsymbol{P}_o = a\boldsymbol{u}_o + b\boldsymbol{E}_o \quad \text{and} \; -\omega^2\boldsymbol{u}_o = a'\boldsymbol{u}_o + b'\boldsymbol{E}.$$

Eliminating $\boldsymbol{u}_o$ from these, we obtain

$$\boldsymbol{P}_o = \left(b + \frac{ab'}{-a - \omega^2}\right)\boldsymbol{E}_o.$$

To derive the dielectric response function $\varepsilon(\omega)$, we use this in the standard formula $\boldsymbol{D}_0 = \varepsilon(\omega)\boldsymbol{E}_o = \varepsilon_o\boldsymbol{E}_o + \boldsymbol{P}_o$, from which we derive

$$\varepsilon(\omega) - \varepsilon_o = b + \frac{ab'}{-a - \omega^2}.$$

Expressing dielectric functions for $\omega = 0$ and ∞ as $\varepsilon(0)$ and $\varepsilon(\infty)$, respectively, we obtain $a' = -\omega_o^2$, $b = \varepsilon(\infty) - \varepsilon(0)$ and $ab'/\omega_o^2 = -b$, so that

$$\varepsilon(\omega) = \varepsilon(\infty) + \frac{\varepsilon(0) - \varepsilon(\infty)}{1 - \dfrac{\omega^2}{\omega_o^2}}. \tag{4.3}$$

Next, we consider that that field $\boldsymbol{E}$ consists of longitudinal and transversal components, i.e. $\boldsymbol{E} = \boldsymbol{E}_L + \boldsymbol{E}_T$, hence $\boldsymbol{u} = \boldsymbol{u}_L + \boldsymbol{u}_T$. For these components, there are spatial relations as follows, i.e.

$$i\boldsymbol{q} \times \boldsymbol{u}_{oL} = 0 \text{ or } \boldsymbol{q} \| \boldsymbol{u}_{oL} \quad \text{and} \quad i\boldsymbol{q}\cdot\boldsymbol{u}_{oT} = 0 \quad \text{or} \quad \boldsymbol{q} \perp \boldsymbol{u}_{oT}$$

for longitudinal and transversal components, respectively.

Keeping the relations, $\boldsymbol{E}_{oL} \| \boldsymbol{u}_{oL}$ and $\boldsymbol{E}_{oT} \| \boldsymbol{u}_{oT}$, the formula div $\boldsymbol{D} = 0$ can be applied to insulators where $\boldsymbol{P}_o = -\varepsilon_o\boldsymbol{E}_o$. Therefore, for longitudinal components, we have

$$\text{div}\,\boldsymbol{u}_L = -\frac{\varepsilon_o + b}{a}\,\text{div}\,\boldsymbol{E}_L \neq 0 \quad \text{and} \; -\omega_L^2\,\boldsymbol{u}_L = a'\,\boldsymbol{u}_L + b'\,\boldsymbol{E}_L.$$

Therefore, the equation of motion for $\boldsymbol{u}_L$ and $\boldsymbol{u}_T$ are expressed as

$$\frac{\partial^2\boldsymbol{u}_L}{\partial t^2} = \left(a' + \frac{ab'}{\varepsilon_o + b}\right)\boldsymbol{u}_L = -\frac{\varepsilon(0)}{\varepsilon(\infty)}\omega_o^2\,\boldsymbol{u}_L \quad \text{and} \quad \frac{\partial^2\boldsymbol{u}_T}{\partial t^2} = a'\,\boldsymbol{u}_T = -\omega_o^2\,\boldsymbol{u}_T,$$

where $\omega_{oL} = \omega_o\sqrt{\frac{\varepsilon(0)}{\varepsilon(\infty)}}$ and $\omega_{oT} = \omega_o$.

Hence, we obtain the Lyddane–Sachs–Teller (LST) relation

$$\frac{\omega_{oL}^2}{\omega_{oT}^2} = \frac{\varepsilon(0)}{\varepsilon(\infty)}. \tag{4.4}$$

Figure 4.1 shows the dielectric response function $\varepsilon(\omega)$ plotted against ω/ω_{oT}, signified by the gap between $\varepsilon(0)$ and $\varepsilon(\infty)$, where a singularity exists at $\omega = \omega_{oT}$ and $\varepsilon(\omega_{oL}) = 0$. Further, we can confirm that the LST relation (4.4) is compatible with Landau's theory. Denoting the transition temperature by T_o in the mean-field accuracy, we write the Curie–Weiss law for the susceptibility as $\chi_<(0) = C/(T - T_o)$ for $T_o < T$ to obtain $\omega_{oL}^2/\omega_{oT}^2 \propto (T_o - T)$, from which we obtain the soft-mode frequency $\tilde{\omega} \propto \sqrt{T_o - T}$ for lattice displacements $\boldsymbol{u}(\boldsymbol{r}, t)$ in mean-field accuracy. Nevertheless, in the vicinity of critical point, soft modes appear to be harmonic, but essentially anharmonic in character.

It is significant to realize that the LST relation (4.4) is derived for homogeneous dielectrics specified by $\boldsymbol{q} = 0$, however both $\boldsymbol{P}_o$ and $\boldsymbol{u}_o$ are proportional to $e^{i\boldsymbol{q}\cdot\boldsymbol{r}}$ that is applicable to inhomogeneous crystals, if $\boldsymbol{q} = \pm \boldsymbol{G}/2$ in particular. Hence, the soft mode cannot be specified by a frequency only, requiring a nonzero wavevector $\boldsymbol{q}$.

4.2 Soft modes in perovskite oxides

Generated by a weak lattice distortion, soft modes were studies in dielectric crystals by applied electric field. In contrast, in perovskite crystals exhibiting zone-boundary transitions, soft modes were detected by neutron inelastic scatterings. In addition, using impurity Fe^{3+} ions in perovskites, soft modes were analyzed from their paramagnetic resonance spectra.

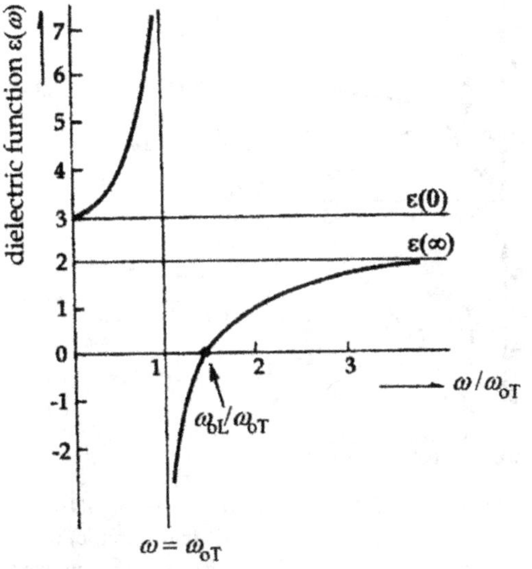

Figure 4.1. Dispersion curve of a dielectric function $\varepsilon(\omega)$. There is a forbidden gap between $\varepsilon(0)$ and $\varepsilon(\infty)$. Reprinted with permission from Springer Nature: from [12]. © 2010.

In any case, order variables (σ_A, σ_P) exhibit soft modes of displacements, offering significant information on structural changes regarding inversion $\boldsymbol{q} \rightleftarrows -\boldsymbol{q}$.

In crystals of $PbTiO_3$[3], softening transversal and longitudinal modes were both reported from (σ_A, σ_P) near the transition point, as shown by well-defined spectra in figure 4.2.

Figure 4.3 [4] summarizes results of magnetic resonance samplings of $SrTiO_3$ and $LaTiO_3$ crystals, showing that a *shear mode* is responsible for softening, as identified from Fe^{3+} spectra. Figure 4.3(*a*) illustrates a model of lattice structure, where the central Ti^{4+} ion is replaced by a paramagnetic Fe^{3+} ion, accompanied by oxygen vacancies at A or B positions indicated in the figure. In magnetic resonance spectra of the Fe^{3+} impurity shown in figure 4.3(*c*) [5], the fine-structure parameter D varies as a function φ of rotation around the [001] axis as expressed by $D \propto \varphi^2$.

4.3 Dynamics of soft modes

It is often convenient to consider a combined object composed of $\boldsymbol{P}$ and $\boldsymbol{u}$, which is called a *condensate*. Dielectric observation of $\boldsymbol{P}$ in laboratories is performed with respect to the relative coordinates. In this regard, we should consider the fluctuation $\Delta\boldsymbol{u}$ in the medium to drive motion of $\boldsymbol{P}$ and $\boldsymbol{u}$ as well.

The function $\Delta\boldsymbol{u}$ for condensates in dielectrics can be expanded with respect to vectors $\boldsymbol{u}_L$ and $\boldsymbol{u}_T$, but actual driving forces are considered as

$$\Delta\boldsymbol{u}_q = \Delta\boldsymbol{u}_{qL} + \Delta\boldsymbol{u}_{qT},$$

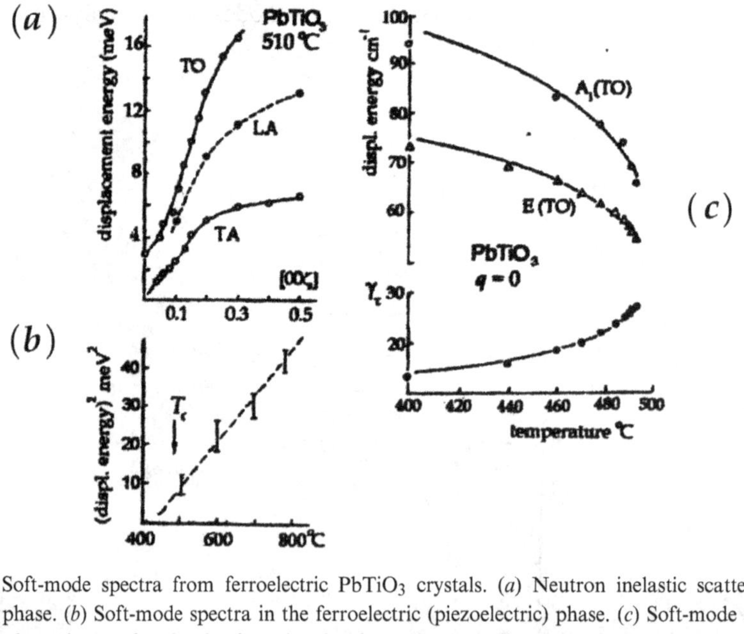

Figure 4.2. Soft-mode spectra from ferroelectric $PbTiO_3$ crystals. (*a*) Neutron inelastic scattering in the paraelectric phase. (*b*) Soft-mode spectra in the ferroelectric (piezoelectric) phase. (*c*) Soft-mode spectra and temperature dependence of γ_τ in the ferroelectric phase. (*a*) and (*b*) reprinted with permission from [3]. Copyright (1970) by the American Physical Society. (*c*) reprinted with permission from [13]. Copyright (1973) by the American Physical Society.

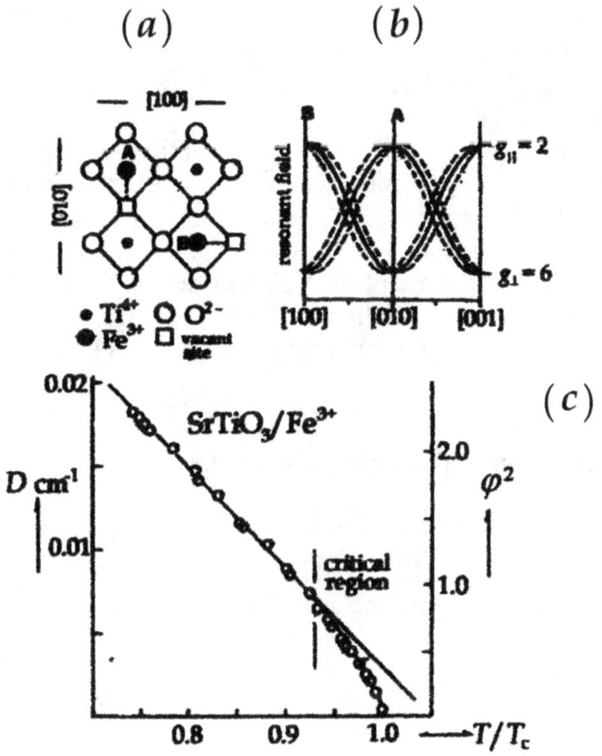

Figure 4.3. Soft-mode spectra of the zone-boundary transitions of $SrTiO_3$ crystals. (*a*) Structural model for zone-boundary fluctuations. (*b*) Variations in a shear angle φ measured from the fine-structure parameter D of Fe^{3+} spectra. (*c*) Soft-mode spectra determined by φ^2 vs temperature. (*c*) reprinted with permission from [14], copyright (1971) and [15], Copyright (1978) by the American Physical Society.

related to the wavevector $\boldsymbol{q}$, where these terms are expanded to

$$\Delta \boldsymbol{u}_{q\mathrm{L}} = V_2 u_{\mathrm{L}}^2 + V_4 u_{\mathrm{L}}^4 + \cdots \quad \text{and} \quad \Delta \boldsymbol{u}_{q\mathrm{T}} = V_2 u_{\mathrm{T}}^2 + V_4 u_{\mathrm{T}}^4 + \cdots,$$

and all other odd terms V_1, V_3, ... are considered for probabilities for lattice relaxation, namely

$$P_{L,T}(\boldsymbol{q}) = 2V_1^2 \langle \boldsymbol{K}' | u_{L,T}(\boldsymbol{q}) | \boldsymbol{K} \rangle^2 \frac{1 - \cos(t\Delta\omega)}{(t\Delta\omega)^2}, \tag{4.5}$$

which should be time-averaged as $\langle P_{L,\,T} \rangle_{\mathrm{t}}$ for damping constant γ_τ and relaxation time τ.

Writing the force as $F_{\mathrm{T}} e^{-i\tilde{\omega}t} = -\nabla(\Delta \boldsymbol{u}_{\mathrm{T}})$ for the transversal energy $\Delta \boldsymbol{u}_{\mathrm{T}} \approx \frac{1}{2} m\omega_{\mathrm{o}}^2 u_{\mathrm{T}}^2$ to drive the soft mode, the equation of motion for u_{T} can be expressed as

$$m\left(\frac{\mathrm{d}^2 u_{\mathrm{T}}}{\mathrm{d}t^2} + \gamma_\tau \frac{\mathrm{d}u_{\mathrm{T}}}{\mathrm{d}t} + \omega_{\mathrm{o}}^2 u_{\mathrm{T}}\right) = F_{\mathrm{T}} e^{-i\tilde{\omega}t}, \tag{4.6}$$

where F_{T} is an internal force, but can represent an external force.

The frequency ω_0 is known as the soft-mode frequency that is characterized by a parabolic temperature dependence. Figure 4.4 shows a typical experimental example obtained from $SrTiO_3$ [5], however a slight discrepancy is evident for the nonlinearity of the soft mode.

Against an effective force, (4.6) has a steady solution for a sufficiently small F_T. Therefore, studies of susceptibilities with F_T provide useful information for the lattice displacement. In experiments, the electric field $E = E_0 e^{-i\tilde{\omega}t}$ with a weak amplitude E_0 is applied to study dielectric soft modes. Although rather unclear in neutron experiments, the presence of weak F_T is evident in results of soft mode study, which are analyzed from susceptibilities.

Assuming $u_T = u_{oT}e^{-i\tilde{\omega}t}$ for a transverse displacement, (4.6) can be written as

$$m\left(-\tilde{\omega}^2 - i\tilde{\omega}\,\gamma_\tau + \omega_o^2\right)u_{oT} = F_T,$$

from which we can define the complex susceptibility

$$\chi(\tilde{\omega}) = \frac{mu_{oT}}{F_T} = \frac{1}{\omega_o^2 - \tilde{\omega}^2 - i\tilde{\omega}\gamma_\tau}. \tag{4.7a}$$

Writing $\chi(\tilde{\omega}) = \chi'(\tilde{\omega}) - i\,\chi''(\omega)$, the real and imaginary parts of (4.7*a*) can be expressed as

$$\chi'(\tilde{\omega}) = \frac{\omega_o^2 - \tilde{\omega}^2}{(\omega_o^2 - \tilde{\omega}^2)^2 + \tilde{\omega}^2\gamma_\tau^2} \quad \text{and} \quad \chi''(\tilde{\omega}) = \frac{\tilde{\omega}\,\gamma_\tau}{(\omega_o^2 - \tilde{\omega}^2)^2 + \tilde{\omega}^2\gamma_\tau^2}, \tag{4.7b}$$

respectively. Figure 4.5 shows these parts plotted against $\tilde{\omega}-\omega_o = \Delta\tilde{\omega}$, which show dispersive and dumbbell-shaped curves, assuming $\gamma_\tau > \frac{|\omega_o^2 - \tilde{\omega}^2|}{\tilde{\omega}^2} \approx 2\Delta\tilde{\omega}$. Otherwise, we have $\chi''(\tilde{\omega}) \sim \frac{1}{\tilde{\omega}\gamma_\tau}$ for $\gamma_\tau > 2\Delta\tilde{\omega}$, exhibiting no peak in an *overdamped* case.

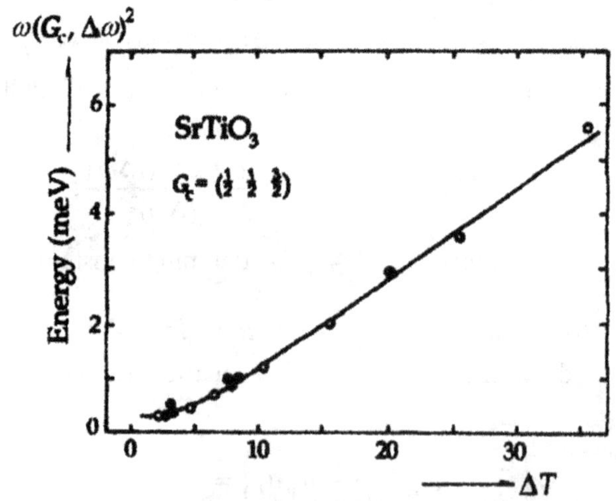

Figure 4.4. A plot of $\omega(G, \Delta\omega)^2$ as *a* function of $T_c - T$ obtained from neutron inelastic scattering experiments of $SrTiO_3$. Reprinted with permission from [15]. Copyright (1978) by the American Physical Society.

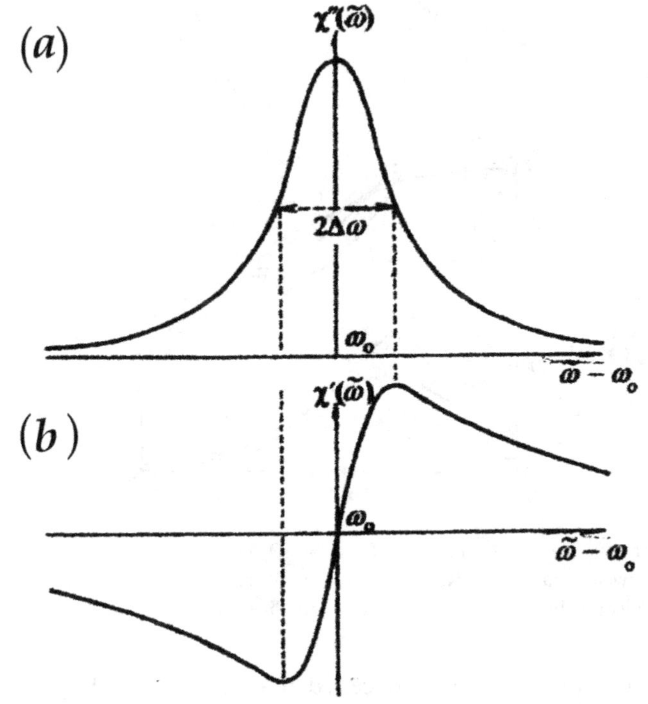

Figure 4.5. A complex susceptibility $\chi(\omega) = \chi'(\omega) - \iota\chi''(\omega)$. (*a*) The imaginary part. (*b*) The real part $\chi'(\omega)$.

Experimentally, however we realize that soft-mode frequency $\tilde{\omega}_0$ is low in the critical region, where equation (4.6) can be simplified as

$$m\left(\gamma_\tau \frac{du_T}{dt} + \omega_o^2 u_T\right) = F_T e^{-i\tilde{\omega}t}. \tag{4.8a}$$

Writing $\frac{\omega_o^2}{\gamma_\tau} = \frac{1}{\tau'}$ and $\frac{F_T}{m\gamma_\tau} = F'$, for convenience, (4.8) is expressed as

$$\frac{du_T}{dt} + \frac{u_T}{\tau'} = F'E^{-i\tilde{\omega}t} \tag{4.8b}$$

where τ' is a *thermal relaxation time*. Characterized by τ', the susceptibility from (4.8*b*) is known as *Debye's relaxation*, that is expressed as

$$\chi_D(\tilde{\omega}) = \frac{1}{-i\tilde{\omega} + \dfrac{1}{\tau'}} = \frac{\tau'}{1 + \tilde{\omega}^2\tau'^2} - \frac{i\tilde{\omega}\tau'}{1 + \tilde{\omega}^2\tau'^2}. \tag{4.9}$$

Figure 4.6 shows examples of Debye's relaxation for soft modes in the paraelectric phase of $BaTiO_3$ crystals [6].

In soft-mode spectra observed by neutron inelastic scatterings at zone boundaries, a sharp absorption peak was found at $\tilde{\omega} \approx 0$, which was unidentified. Nevertheless, it was analyzed for possible significant findings. In any case, it can be analyzed as

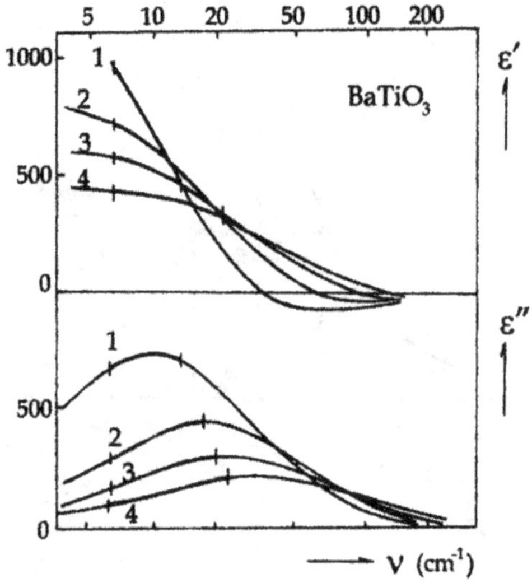

Figure 4.6. Dielectric spectra of $\varepsilon'(\omega)$ and $\varepsilon''(\omega)$ from $BaTiO_3$ measured with the backward-wave technique. Curves 1, 2, 3 and 4 were obtained at 474, 535, 585 and 667 K, respectively. Reprinted from [6] by permission of the publisher (Taylor & Francis Ltd, http://www.tandfonline.com).

relaxation to unknown origin, and called a *central peak*. Figure 4.7 illustrates examples of central peaks in soft-mode spectra, observed in $SrTiO_3$ and $KMnF_3$ [7].

For the central peak described by v, we express Debye's equation as

$$\frac{\mathrm{d}v}{\mathrm{d}t} + \frac{v}{\bar{\tau}} = \bar{F}e^{-i\tilde{\omega}t}, \tag{4.10}$$

which is combined with the equation for u_T as

$$\frac{\mathrm{d}^2u_\mathrm{T}}{\mathrm{d}t^2} + \gamma_\tau\frac{\mathrm{d}u_\mathrm{T}}{\mathrm{d}t} + c\bar{\gamma}\frac{\mathrm{d}v}{\mathrm{d}t} + \omega_\mathrm{o}^2u_\mathrm{T} = \frac{F}{m}e^{-i\tilde{\omega}t},$$

where γ and c are the coupling constant between u_T and v.

The susceptibility can then be derived as

$$\chi(\tilde{\omega}) = \frac{mu_{\mathrm{oT}}}{F} = \left(\omega_\mathrm{o}^2 - \tilde{\omega}^2 - i\tilde{\omega}\gamma_\tau - ic\bar{\gamma}\bar{F}\frac{\tilde{\omega}\tau}{1 - i\tilde{\omega}\tau}\right)^{-1},$$

which is identical to (4.7), if $c = 0$. In this case, the displacement v of (4.10) is obviously independent from u_T.

4.4 Soft-mode frequency in modulated crystals

Temperature-dependent frequencies of soft modes are typically expressed as $\tilde{\omega}_\mathrm{o}^2 \propto T_c - T$ in figure 4.1, which is adequate for spectroscopic measurements, except for the critical region. However, owing to the fluctuating wavevector $\boldsymbol{q}$, the volume change ΔV is unavoidable but neglected because of $\Delta V \approx 0$.

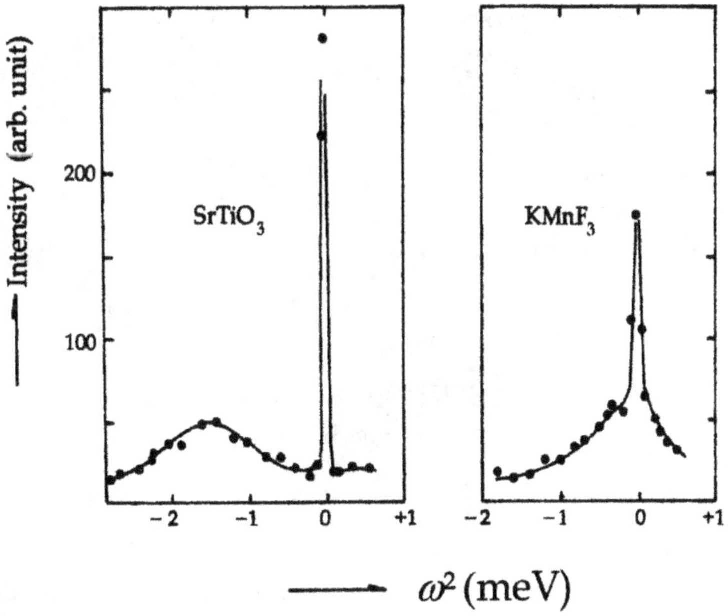

Figure 4.7. Soft-mode spectra from $SrTiO_3$ and $KMnF_3$, including sharp central peaks at $\omega = 0$. Reprinted figure with permission from [7]. Copyright (1972) by the American Physical Society.

Cowley discussed that it arises from the response function to the fluctuation energy $\Delta \boldsymbol{u}$ in the lattice. Following his theory [8], such a temperature dependence is described in this section, as related to phonon scatterings in the thermodynamic situation.

In dielectric studies, the displacement u_T has a direction parallel to the applied field $\boldsymbol{E}$. However, if $\boldsymbol{E} = 0$, the direction of u_T is arbitrary in equilibrium crystals, so we write $u_T = u_{Ty} + iu_{Tz}$ with respect to the longitudinal x-axis. Then, the fluctuating Δu_C is given by

$$\Delta u_C = V_3 u_L^2 (u_{Ty} \pm iu_{Tz}) + V_4 u_L^2 \left(u_{Ty}^2 + u_{Tz}^2\right) \tag{4.11}$$

that is fourth-order terms in the mean-field approximation. Therefore, for the transversal u_T, writing a complex damping constant $\gamma = \Gamma - i\Phi$, the steady transversal displacement is expressed as

$$u_T(\boldsymbol{q}, \Delta\omega) = \frac{F_{\pm}/m}{-\omega(\boldsymbol{q})^2 + \omega_0^2 - i\,\omega(\boldsymbol{q})(\Gamma - i\,\Phi)}.$$

The corresponding frequency is shifted from ω_0 to ω derived from

$$\omega(\boldsymbol{q}, \Delta\omega)^2 = \omega_0^2 + \omega(\boldsymbol{q})\Phi(\boldsymbol{q}, \Delta\omega), \tag{4.12}$$

where $\Phi(\boldsymbol{q}, \Delta\omega)$ can be calculated with Δu_C of (4.11), but neglecting $\Phi_0 = \frac{\partial\omega(\boldsymbol{q})}{\partial V}\Delta V$ for $\Delta V \approx 0$.

Therefore, writing $\Phi(\boldsymbol{q}, \Delta\omega) = \Phi_1(\boldsymbol{q}) + \Phi_2(\boldsymbol{q}, \Delta\omega)$, where

$$\Phi_1(\boldsymbol{q}) = \frac{\hbar}{N\omega(\boldsymbol{q})}\frac{(2n+1)V_4 u_L^2}{2\omega'}\langle -\boldsymbol{K}, \boldsymbol{K}|u_T^2|\boldsymbol{K}', -\boldsymbol{K}'\rangle$$

for phonon scatterings $(K, K') \rightleftarrows (-K, -K')$ at $\frac{\hbar\omega'}{2}(n+1) = k_B T$. Here, n is phonon number, and N signifies the total, so that $\Phi_1(\boldsymbol{q}) \propto T$.

Further calculations yield the relation

$$\Gamma(\boldsymbol{q}, \Delta\omega) = \Phi_2(\boldsymbol{q}, \Delta\omega) = \frac{\pi\hbar}{32N\omega(\boldsymbol{q})}\sum_{1,2}\frac{V_4 u_L^2\langle \boldsymbol{q}|u_T|\boldsymbol{K}_1, \boldsymbol{K}_2\rangle^2}{\omega_1\omega_2}$$
$$\times (n_1 + n_2 + 1)\{-\delta(\omega + \omega_1 + \omega_2) + \delta(\omega - \omega_1 - \omega_2)\}$$
$$\times (n_1 - n_2)\{-\delta(\omega - \omega_1 + \omega_2) + \delta(\omega + \omega_1 - \omega_2)\}$$

to define temperatures T_1 and T_2 by $\hbar\omega(n_1 + n_2 + 1) = k_B T_1$ and $\hbar\omega(n_1 - n_2) = k_B T_2$, and the damping factor $\Gamma(\boldsymbol{q}, \Delta\omega)$ indicates that the phono energy $\hbar\omega(2n_2 + 1) = k_B(T_1 - T_2)$ is transferred to surroundings.

The soft-mode frequency appears to converge to zero, while prohibited by transitional anomaly, depending on the timescale of bifurcation. Experimentally, $\omega(\boldsymbol{q})$ is signified by $\Delta\omega$ and obscured near $\boldsymbol{q} = 0$ in zone-center transitions, for which we write a relation

$$\tilde{\omega}^2 = \omega(0, \Delta\omega)^2 = \omega_o^2 + \Phi_1(0)(\omega_o \pm \Delta\omega) \approx 0 \tag{4.13a}$$

to derive $\omega_o \sim \pm\Delta\omega/2$, confirming Landau's relation $\tilde{\omega}^2 = A'(T_o - T) \approx 0$ near T_o, where A' is a constant.

On the other hand, zone-boundary transitions at $\boldsymbol{q} = \pm \boldsymbol{G}/2$, (4.13$a$) should be modified as

$$\tilde{\omega}(\boldsymbol{q}, \Delta\omega)^2 = \tilde{\omega}(0, \Delta\omega)^2 + \kappa^2\boldsymbol{q}^2 + \cdots. \tag{4.13b}$$

Accordingly, the following expressions are provided for soft modes for $\boldsymbol{q} \neq 0$ at Brillouin-zone boundaries. Namely,

$$\tilde{\omega}^2 \approx \omega(\boldsymbol{q}, \Delta\omega)^2 = A'(T - T_o) + \kappa^2\boldsymbol{q}^2 \quad \text{for} \quad T > T_o \tag{4.14a}$$

and

$$\tilde{\omega}^2 = \omega(\boldsymbol{q}, \Delta\omega)^2 \approx A'(T_o - T) + \kappa^2\boldsymbol{q}^2 \quad \text{for} \quad T < T_o. \tag{4.14b}$$

Soft-mode frequencies from binary systems show mostly a parabolic temperature dependence as described by $(T_c - T)^{1/2}$ for $T < T_c$, analyzed for an exponential form $(T_c - T)^\beta$ to determine the critical exponent β by experiments. However, experimental results indicate that the value of β depends on the dimensionality of crystals, as predicted by the *scaling theory* [9], while close to parabolic in three dimensions.

4.5 Optical studies on symmetry changes at critical temperature

4.5.1 Cochran's model of ferroelectric transitions

Soft modes were observed a long time ago, as Cochran [10] discussed ferroelectric transitions in 1960. He showed that such softening modes occur in consequent on counteracting short- and long-range interactions in crystals.

In dielectric crystals composed by positive and negative ions ($+e$, $-e$), the reduced mass $\mu = \frac{m_+ m_-}{m_+ + m_-}$ of an ionic pair (m_+, m_-) is displaced by distance $u = u_+ - u_-$ to create a dipole moment $p = eu$ at each site of the ionic lattice, which is responsible for macroscopic polarization P parallel along with the longitudinal direction.

Considering for the transversal lattice displacement $\boldsymbol{u}_\mathrm{T}$ to be perpendicular to the electric polarization $\boldsymbol{P}$ that is parallel to the longitudinal $\boldsymbol{u}_\mathrm{L}$, Cochran wrote the equations

$$m_+ \frac{\mathrm{d}^2 u_{T+}}{\mathrm{d}t^2} = -C(u_{T+} - u_{\mathrm{T}-}) + e\left(E + \frac{P}{3\varepsilon_\mathrm{o}}\right) \quad \text{and}$$

$$m_- \frac{\mathrm{d}^2 u_{\mathrm{T}-}}{\mathrm{d}t^2} = -C(u_{\mathrm{T}-} - u_{T+}) - e\left(E + \frac{P}{3\varepsilon_\mathrm{o}}\right),$$

to derive

$$\mu \frac{\mathrm{d}^2 P}{\mathrm{d}t^2} = \frac{e^2}{V}\left(E + \frac{P}{3\varepsilon_\mathrm{o}}\right) - CP. \tag{4.15}$$

For the applied electric field $E = E_\mathrm{o} e^{i\tilde{\omega}_\mathrm{T} t}$, we write the polarization $P = P_\mathrm{o} e^{i\tilde{\omega}_\mathrm{T} t}$ to obtain the susceptibility

$$\chi(\tilde{\omega}_\mathrm{T}) = \frac{P}{\varepsilon_\mathrm{o} E} = \frac{e^2/\varepsilon_\mathrm{o} V}{C - (e^2/\varepsilon_\mathrm{o} V) - \mu\tilde{\omega}_\mathrm{T}^2}, \tag{4.16}$$

showing a singularity for a soft mode at

$$\mu\tilde{\omega}_\mathrm{T}^2 = C - (e^2/\varepsilon_\mathrm{o} V) = 0, \text{ if } C = e^2/\varepsilon_\mathrm{o} V.$$

On the other hand, for a longitudinal polarization $P = \frac{e}{V}(u_{\mathrm{L}+} - u_{\mathrm{L}-})$, we have that equation (4.15) is expressed as

$$\mu \frac{\mathrm{d}^2 P}{\mathrm{d}t^2} = -\frac{4e^2 P}{3\varepsilon_\mathrm{o} V} - CP \quad \text{for} \quad E = 0,$$

exhibiting no singularity for $\tilde{\omega}_\mathrm{L}$.

4.5.2 Symmetry change at transition temperatures

Symmetry change is a visible signature of structural phase transitions, which needs to be substantiated for the theory to be correct. Among crystals in many types, soft modes observed in TSCC and perovskites have provided supporting results of structural changes at T_c.

Figure 4.8 shows that soft modes in TSCC are identified as in B_{2u} and A_1 symmetries for $T > T_c$ and $T < T_c$, respectively, indicating that their directions of propagation are perpendicular to each other. Therefore, these modes are considered as driven by potential energies in corresponding symmetries. Expressing these by

$$\Delta \boldsymbol{u}_{B_{2u}} = \frac{A}{2}u_x^2 + \frac{B}{4}u_x^4 \quad \text{and} \quad \Delta \boldsymbol{u}_{A_1} = \frac{C}{2}u_y^2 + \frac{D}{4}u_y^4,$$

where the coordinates are set for the x- and y-axes to represent longitudinal and transversal directions, respectively; and A, B, C and D are constants.

Corresponding to a transform between B_{2u} and A_1 phonon modes, we consider that the lattice displacement is converted between u_x and u_y', which can be expressed by a linear combination $u' = \mathbf{c}_1\boldsymbol{u}_x + \mathbf{c}_2\boldsymbol{u}_y'$ and $c_1^2 + c_2^2 = 1$. By this, the change $B_{2u} \rightarrow A_1$ can be specified by $(c_1 = 1,\ c_2 = 0) \rightarrow (c_1 = 0,\ c_2 = 1)$, and described by

$$\lim \Delta \boldsymbol{u}_{\text{trans}} \rightarrow -\frac{A}{4}u'^2_y = \Delta \boldsymbol{u}_{A_1}, \text{ ignoring } D \approx 0.$$

Soft-mode frequencies curves in figure 4.8 [11] are dominated by parabolic temperature dependence on both sides of T_c. Experimentally, it is noted that observed leading terms of $\Delta \boldsymbol{u}$ are parabolic terms $\frac{A}{2}u^2$ and $\frac{A}{4}u^2$ for $T < T_c$ and $T > T_c$, respectively

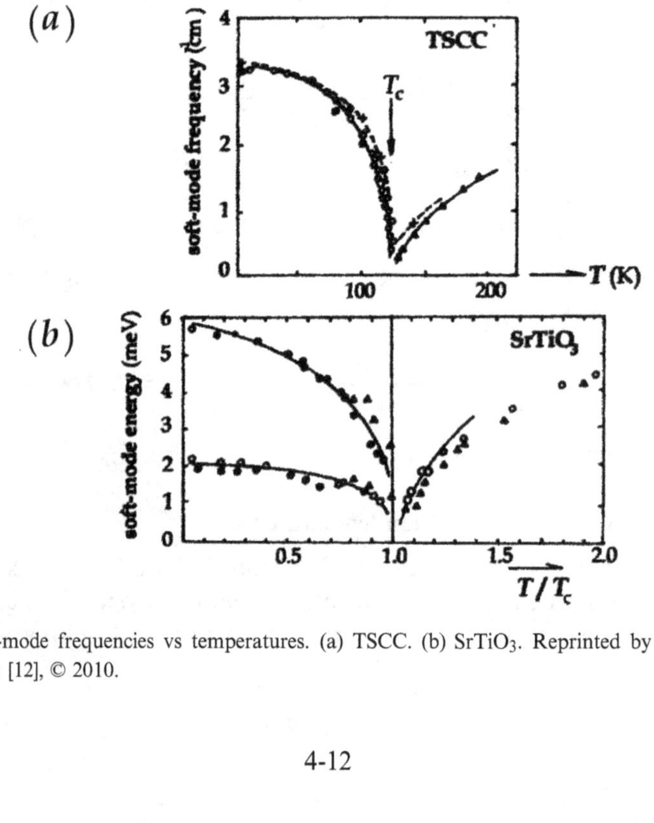

Figure 4.8. Soft-mode frequencies vs temperatures. (a) TSCC. (b) $SrTiO_3$. Reprinted by permission from Springer Nature: [12], © 2010.

Exercises

1. According to section 4.3, the soft mode originates from adiabatic anharmonic potential V_4, which is temperature dependent. This indicates that V_4 cannot be a continuous perturbation, where the discontinuity may cause energy transfer to the lattice. Discuss this issue of singularity qualitatively.
2. Soft modes represent the response from the lattice to the transverse component of vector waves, whereas scattering and magnetic resonance studies deal with the longitudinal component of pseudospins. Discuss this problem with respect to two components of nonlinear waves.

References

[1] Elliott R J and Gibson A F 1975 *An Introduction to Solid State Physics and its Applications* (London: McMillan)
[2] Born M and Huang K 1968 *Dynamical Theory of Crystal Lattices* (London: Oxford University Press)
[3] Shirane G, Axe J D, Harada J and Remeika J P 1970 *Phys. Rev.* B **2** 155
[4] Kirkpatrick E S, Müller K A and Rubins R S 1964 *Phys. Rev.* **135** A86
Currat R, Müller K A, Berlinger W and Desnoyer R 1978 *Phys. Rev.* B **17** 2938
[5] Unoki H and Sakudo T 1967 *J. Phys. Soc. Jpn.* **23** 546
[6] Petzelt J, Kozlov G V and Volkov V V 1987 *Ferroelectrics* **73** 101
[7] Shapiro S M, Axe J D and Riste T 1972 *Phys. Rev.* B **6** 4332
[8] Cowley A 1968 *Prog. Phys.* **31** 123
[9] Stanley H E 1971 *Introduction to Phase Transitions and Critical Phenomena* (New York: Oxford University Press)
[10] Cochran W 1973 *The Dynamics of Atoms in Crystals* (London: Edward Arnold)
[11] Scott J I 1983 *Raman Spectroscopy of Structural Phase Transitions in Light Scatterings near Phase Transitions* ed H Z Cummins and A P Levanyak (Amsterdam: North Holland)
[12] Fujimoto M 2010 *Thermodynamics of Crystalline States* (Berlin: Springer)
[13] Burns G and Scott B A 1973 *Phys. Rev.* B **7** 3088
[14] Muller K A and Berlinger W 1971 *Phys. Rev. Lett.* **26** 13
[15] Currat R, Muller K A, Berlinger W and Desnoyer R 1978 *Phys. Rev.* B **17** 2937

IOP Publishing

Introduction to the Mathematical Physics of Nonlinear Waves
(Second Edition)

Minoru Fujimoto

Chapter 5

Nonlinearity development in crystals: Korteweg–deVries' equation for collective order variables and the complex potential

By mathematical solution of solitons, we can deal with correlations among order variables in media with respect to *Newton's action–reaction principle*, yielding useful information about responding surroundings. Represented by the adiabatic potential in thermodynamics, the *complex soliton potential* provides the properties of lattice in better detail than in mean-field accuracy, constituting an essential property of nonlinearity, while from the Korteweg–deVries equation, we derive the thermodynamic potential for the crystal symmetry to be stabilized.

5.1 The Korteweg–deVries equation

5.1.1 Timescale for developing nonlinearity

Models in mechanics cannot be adapted for general use for dynamical applications in microscopic systems, unless signified by eigenstates of the Hamiltonian. In principle, the model should be supported for dynamical systems to be used in thermal equilibrium with surroundings. In other words, the system should be *conservative* in character.

The pressure from surroundings, for example, is traditionally assumed to be constant for thermodynamic equilibrium. On the other hand, the perturbing pressure gradient related to $p_1' = \Delta p$ in section 2.7 represents a force by the reacting surrounding, whereby a potential V_1 is defined as related with $p_1' \propto V_1$, corresponding to an *adiabatic potential* in thermodynamics. In there, we found that $V_1(\xi)$ is in phase with $\rho_1'(\xi)$ and $v_1'(\xi)$ in the first-order approximation, moving together at the speed $v_1'(\xi)$; and $\rho_1'(\xi)$ scatteredby $V_1'(\xi)$. Arising from $V_1'(\xi)$, the nonlinearity is developed as a function of τ that is normally time t, but representing temperature T of surroundings. In the latter case, it is essential to consider *phonons* in the Hamiltonian to include thermal interaction, leaving t as usual for time variation ΔT in $V_1'(\xi)$.

In nonlinear dynamics, the development of nonlinearity can be discussed by a wavefunction $\psi(\boldsymbol{r}, \tau)$ for the responsible order variable $\sigma(\boldsymbol{r}, \tau)$, where τ represents time for dynamical discussion, while *temperature* T is for thermodynamics; although any parameter for development is acceptable theoretically.

We consider the development equation written for the wavefunction $\psi(\boldsymbol{r}, \tau)$, which is given as

$$\frac{\partial \psi}{\partial \tau} = \mathcal{B}\psi, \tag{5.1}$$

where $\mathcal{B}$ is called a *development operator*, and τ a development parameter, i.e. time or temperature.

We consider that such a potential as $V_1(\boldsymbol{r}, \tau)$ is developed by a structural change in crystals with respect to the relation between order variables in the lattice, as a subject to the *action–reaction principle*. Namely, if $+V_1(\boldsymbol{r}, \tau)$ is developed in crystals, $-V_1(\boldsymbol{r}, \tau)$ takes place to pin $\boldsymbol{\sigma}(\boldsymbol{r}, \tau)$ to the lattice.

5.1.2 The Korteweg–deVries equation

Expanding the accuracy k^2 of adiabatic approximation, the Hamiltonian is expressed as

$$\mathcal{H} = \mathcal{H}_o + \mathcal{H}_1 k^2 + \mathcal{H}_2 k^4 + \cdots,$$

where k^2 is a parameter related to the reduced/total mass ratio in crystals. We assume that $k^2\mathcal{H}_1$ represents $\mathcal{H}$ in adiabatic approximation.

In the presence of a Weiss' potential V_1, we express Schrödinger's equation

$$\mathcal{H}_1\psi = (\mathcal{D}^2 - V_1)\psi = \varepsilon_1\psi \tag{5.2}$$

for the eigenvalue ε_1, corresponding to a steady state at constant τ, and $\mathcal{D} = \nabla_\xi$ is a differential operator. Therefore, $\frac{\partial \varepsilon_1}{\partial \tau} = 0$ should be applied to thermal equilibrium.

This relation (3.2) can be confirmed as follows. Namely,

$$\frac{\partial}{\partial \tau}(\mathcal{H}_1\psi) = \frac{\partial \mathcal{H}_1}{\partial \tau}\psi + \mathcal{H}_1\frac{\partial \psi}{\partial \tau} = -\frac{\partial V_1}{\partial \tau}\psi + \mathcal{H}_1\mathcal{B}\psi$$

and

$$\frac{\partial}{\partial \tau}(\varepsilon_1\psi) = \frac{\partial \varepsilon_1}{\partial \tau}\psi + \varepsilon_1\frac{\partial \psi}{\partial \tau} = \varepsilon_1(\mathcal{B}\psi) = \mathcal{B}(\mathcal{H}_1\psi),$$

where these expressions are identical to each other. Hence, we obtain

$$\frac{\partial V_1}{\partial \tau}\psi - (\mathcal{H}_1\mathcal{B} - \mathcal{B}\mathcal{H}_1)\psi = 0 \quad \text{or} \quad \left(-\frac{\partial V_1}{\partial \tau} + [\mathcal{H}_1, \mathcal{B}]\right)\psi = 0. \tag{5.3}$$

For the operator $\mathcal{B}$, we consider the expansion

$$\mathcal{B} = a_1\mathcal{D} + a_2\mathcal{D}^2 + a_3\mathcal{D}^3 + \ldots\ldots,$$

where a_1, a_2, a_3, ... are all functions of ξ.

If the development is determined only by the first term, i.e. $\mathcal{B} = \mathcal{B}_1 = a_1\mathcal{D}$, equation (5.3) can be expressed by $\frac{\partial\psi}{\partial\tau} = a_1\frac{\partial\psi}{\partial\xi}$, hence $\psi(\xi - a_1\tau)$ is an harmonic wave at a phase velocity a_1. Assuming $\mathcal{B}_2 = a_2\mathcal{D}^2$, the propagation also remains as linear in the phase $\xi - a_2\tau$.

On the other hand, considering $\mathcal{B}_3 = a_3\mathcal{D}^3 + a_1\mathcal{D} + a_o$, these constants can be determined from $V_1(\xi)$ as functions of ξ.

The commutator in (5.3) can be calculated as shown below.

$$[\mathcal{H}_1, \mathcal{B}_3] = \left(2\frac{\partial a_1}{\partial\xi} + 3a_3\frac{\partial V_1}{\partial\xi}\right)\mathcal{D}^2\psi + \left(\frac{\partial^2 a_1}{\partial\xi^2} + 2\frac{\partial a_o}{\partial\xi} + 3\frac{\partial^2 V_1}{\partial\xi^2}\right)\mathcal{D}\psi$$
$$+\left(\frac{\partial^2 a_o}{\partial\xi^2} + a_3\frac{\partial^3 V_1}{\partial\xi^3} + a_3\frac{\partial V_1}{\partial\xi}\right)\psi,$$

where should be zero for $\frac{\partial V_1}{\partial\tau} = 0$, but the coefficient of $\mathcal{D}^2\psi$ should be zero, and $\mathcal{D}\psi = 0$. Therefore, we have two relations

$$2\frac{\partial a_1}{\partial\xi} + 3a_3\frac{\partial V_1}{\partial\xi} = 0 \quad \text{and} \quad \frac{\partial^2 a_1}{\partial\xi^2} + 2\frac{\partial a_o}{\partial\xi} + 3a_3\frac{\partial^2 V_1}{\partial\xi^2} = 0,$$

which can be integrated as

$$a_1 = -\frac{3}{2}a_3V_1 + c \quad \text{and} \quad a_o = -\frac{3}{4}a_3\frac{\partial V_1}{\partial\xi} + c',$$

where c and c' are integration constants. In this case, we have

$$[\mathcal{H}_1, \mathcal{B}_3]\psi = \left\{\frac{a_3}{4}\left(\frac{\partial^3 V_1}{\partial\xi^3} - 6V_1\frac{\partial V_1}{\partial\xi}\right) + c\frac{\partial V_1}{\partial\xi}\right\}\psi.$$

Accordingly, equation (5.3) can be expressed as

$$\frac{a_3}{4}\left\{\frac{\partial^3 V_1}{\partial\xi^3} - (6V_1 + c)\frac{\partial V_1}{\partial\xi} - \frac{\partial V_1}{\partial\tau}\right\}\psi = 0.$$

Choosing $a_3 = -4$ and $c = 0$, this equation can be written as

$$\frac{\partial V_1}{\partial\tau} - 6V_1\frac{\partial V_1}{\partial\xi} + \frac{\partial^3 V_1}{\partial\xi^3} = 0, \tag{5.4}$$

which is the standard form of the *Korteweg–deVries equation.*

It is noted that for the potential $V_1(\xi)$ determined by (5.4), the function $\psi(\xi, \tau)$ is a developing solution of

$$\frac{\partial\psi}{\partial\tau} = \mathcal{B}_3\psi = \left(-4\mathcal{D}^3 + 6V_1\mathcal{D} + 3\frac{\partial V_1}{\partial\xi}\right)\psi \quad \text{and} \quad \frac{\partial\varepsilon_1}{\partial\tau} = 0. \tag{5.5}$$

At this point, it is significant to assume $c = 0$ to derive (5.4), because the system is not conservative if $c \neq 0$.

In canonical systems, the dynamical phase $\xi - a_1\tau$ specifies the speed of propagation $a_1 = 6V_1$ in the presence of lattice potential k^2V_1, requiring $c = 0$. In section 2.6, we had the density $\rho_1'(\xi) = \psi^*(\xi)\psi(\xi)$ moving together at speed $6V_1$. Furthermore, the change ΔV_1 dueto $\Delta\tau$ should be attributed to ΔT and Δp for isothermal and adiabatic transition, respectively.

5.2 Thermal solution for the Weiss potential

It is significant that the Korteweg–deVries (KdV in short) equation has an analytical solution for steady nonlinear phenomena in thermodynamics.

In a potential V_1 pinned in a Galilean system, it is convenient to define a phase variable $\phi = x - v\tau$ to keep the conventional speed v and specific τ for the present mathematical argument. With that specification, we have the relation

$$\frac{\partial V_1}{\partial\tau} - v\frac{\partial V_1}{\partial x} = 0, \tag{5.6}$$

solving the KdV equation for $V_1(x - v\tau)$ to discuss its space-time properties. With this definition, equation (5.4) can be specified by the phase zero at $x = v\tau$, where (5.4) is expressed as

$$v\frac{\partial V_1}{\partial x} - 3\frac{\partial(V_1^2)}{\partial x} + \frac{\partial}{\partial x}\left(\frac{\partial^2 V_1}{\partial x^2}\right) = 0,$$

where the spatial phase is x, as supported by Galilean invariance, then integrated as

$$\frac{\mathrm{d}^2 V_1}{\mathrm{d}x^2} = 3V_1^2 - vV_1 + a, \qquad \text{where } a \text{ is a constant.}$$

Multiplying by $\frac{\mathrm{d}V_1}{\mathrm{d}x}$ on both sides, we obtain

$$\frac{1}{2}\left(\frac{\mathrm{d}V_1}{\mathrm{d}x}\right)^2 = V_1^3 - \frac{v}{2}V_1^2 + aV_1 + b,$$

where b is another integration constant. The right side is an algebraic expression of the 3rd order with respect to V_1, which can generally be factorized as

$$\left(\frac{\mathrm{d}V_1}{\mathrm{d}x}\right)^2 = -2(V_1 - V_{1A})(V_1 - V_{1B})(V_1 - V_{1C}), \tag{5.7}$$

where V_{1A}, V_{1B} and V_{1C} are three roots of the algebraic equation $\left(\frac{\mathrm{d}V_1}{\mathrm{d}x}\right)^2 = 0$.

Curves in figure 5.1 show two cases, one specified by two roots at A and B, the other one only at C. A and B in the former join C in the latter, suggesting that a

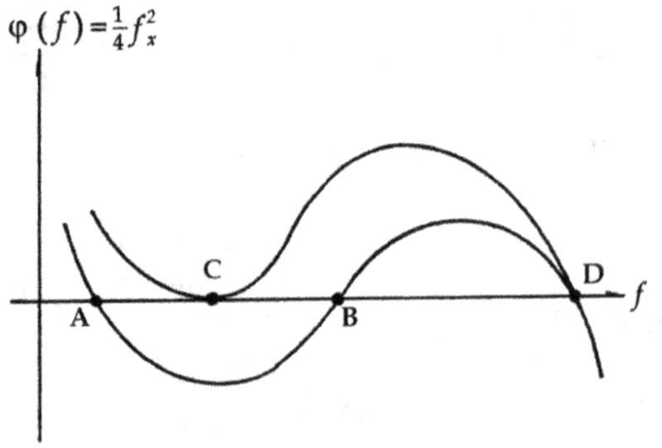

Figure 5.1. Thermodynamic Weiss potential: its prediction, indicating a negative potential between A and B, critical at C.

lattice potential exists between A and B, while point C represent the critical temperature T_c.

Such a lattice potential $V(\phi)$ can be defined as proportional to $\left(\frac{dV_1}{dx}\right)^2$ and exhibits oscillation between A and B in analogy to figure 1.3 in chapter 1. Hence, considering another function $g = V_1 - V_{1C}$ for the oscillatory case, we write

$$\left(\frac{dg}{dx}\right)^2 = 2g(V_{1C} - V_{1A} - g)(V_{1C} - V_{1B} - g).$$

Further, introducing another phase variable ζ by writing $g = (V_{1C} - V_{1B})\zeta^2$, the above expression can be changed to

$$\left(\frac{d\zeta}{dx}\right)^2 = \frac{V_{1C} - V_{1A}}{2}(1 - \zeta^2)(1 - \kappa^2\zeta^2), \quad \text{where} \quad \kappa^2 = \frac{V_{1C} - V_{1B}}{V_{1C} - V_{1A}}.$$

Finally, to introduce the new phase $\phi = \sqrt{V_{1C} - V_{1A}}\,\zeta$, the above can be re-expressed for

$$2\left(\frac{d\zeta}{d\phi}\right)^2 = (1 - \zeta^2)(1 - \kappa^2\zeta^2)$$

to be integrated as

$$\frac{\phi_1}{\sqrt{2}} = \int_0^{\zeta_1} \frac{d\zeta}{\sqrt{(1 - \zeta^2)(1 - \kappa^2\zeta^2)}}. \tag{5.8a}$$

Here the phase ϕ_1 is determined by the upper limit ζ_1 of the elliptic integral defined in chapter 1. The reverse function of (5.8*a*) is known as Jacobi's sn-function

$$\zeta_1 = \mathrm{sn}\left(\frac{\phi_1}{\sqrt{2}}, \kappa\right). \tag{5.8b}$$

The *lattice potential* $V(\phi)$ can therefore be expressed as

$$V(\phi) = V_{1C} - (V_{1C} - V_{1B})\mathrm{sn}^2\left(\sqrt{V_{1C} - V_{1A}}\ \phi, \kappa\right) \quad \text{for} \quad 0 < \kappa < 1,$$

where the sn-function is periodic with the period

$$2K(\kappa) = \int_0^1 \frac{d\zeta}{\sqrt{(1 - \zeta^2)(1 - \kappa^2\zeta^2)}}.$$

On the other hand, if V_{1A} become close to V_{1B} as A → B, no such lattice potential is possible thermodynamically, as indicated by $2K(1) \to \infty$. Thermodynamic stability in the lattice is determined by $\kappa \to 0$, where $V_\perp(\phi)$ is given by

$$V(\phi) = V_{1C} + (V_{1C} - V_{1B})\mathrm{sech}^2\left(\sqrt{V_{1C} - V_{1A}}\ \phi\right), \tag{5.9c}$$

showing the *pulse-shape* with height $|V_{1C} - V_{1B}|$ and effective phase $\sqrt{V_{1C} - V_{1A}}\ \phi$.

Experimentally, the potential $V(\phi)$ is known as temperature dependent, appearing with a weak amplitude after phase transitions, where the frequency is relatively high, changing from pulse to sinusoidal as mathematically predicted from (5.9). The following potential should be applied to this situation. That is

$$V(\phi) = V_{1C} + (V_{1C} - V_{1B}) \sin^2\left(\sqrt{V_{1C} - V_{1A}}\ \phi\right). \tag{5.9d}$$

The parameter κ of the elliptic function is a mathematical option, but $\kappa = 1$ provides access to the symmetry to characterize crystals in thermal equilibrium. Judging from the definition of $V(\phi)$ from (5.7), the relation with the variable V_1 was somewhat complicated, but the potential undoubtedly exists in nature as a real one, representing *the feature of the Weiss field*, as well as representing the *complex lattice* in mean-field accuracy. Therefore, in this book we shall call $V(\phi)$ the *Weiss potential.*

We note that the origin for confusion arises from two-component waves in nonlinear processes, corresponding mathematically to a *complex potential.* Hence, this issue of nonlinear motion will be explained in the following sections.

5.3 Condensate pinning by the Weiss potential

The Weiss potential is essential for the nonlinearity of order variables in the lattice, in a similar manner to the traditional concept of condensates of composites and solvents. In fact, it is a matter of relative coordinates between variables and the lattice. In this regard, $-V(\phi)$ acts on $\boldsymbol{\sigma}(\phi)$, whereas $+V(\phi)$ is responsible for the lattice stability; but in the center-of-mass system both are canceled out and indetectable. Nevertheless, the mathematics of (5.9*c*) has been a popular subject since the mid-19th century, as discussed in chapter 2. Therefore, referring to section 2.3.2, the problem for the problem of pinning by $-V(\phi)$ continues.

For the pinning potential, we consider a negative potential written as $V(x) = -V_0\ \mathrm{sech}^2\frac{x}{d}$, where $V_0 > 0$ and the width is $2d$ for a negative state $-E_1$. No reflection of $\sigma_1 \propto e^{\pm kx}$ is possible, hence we use position x, instead of the phase $x - v\,\tau$, to determine the eigenstates of σ_1 in the potential $V(x)$. Accordingly, we started with reduced potential $v = 2md^2V_o/\hbar^2$ and eigenvalue $\varepsilon = 2md^2E_1/\hbar^2$ to express the wave equation of $\sigma_1(x)$ as shown below. Namely,

$$\frac{d^2\sigma_1}{dz^2} + (\varepsilon + v\ \mathrm{sech}^2 z)\sigma_1 = 0 \quad \text{where} \quad z = \frac{x}{d}, \tag{5.10a}$$

and after tedious calculation, resulting in

$$v = n(n+1) \qquad \text{and} \qquad \varepsilon = n, n-1, n-2, \ldots, 2, 1. \tag{5.10b}$$

Therefore, the number n can specify the pinning potential $V(\phi)$ at discrete levels where two $\boldsymbol{\sigma}_1$ and $\boldsymbol{\sigma}_2$ are symmetrically combined together in parallel form $\sigma_{\perp 1}||\sigma_{\perp 2}$ at a critical point T_n, so that the pinning potential is expressed by $V(\sigma_\perp^2)$. Notice that a symmetric combination is consequent on *boson statistics* of solitons. In any case, we consider that (5.10*b*) represents the basic property of the *Weiss potential* $V(\phi_c, T_n)$ at T_c.

5.4 Nonlinear waves and complex lattice potentials

5.4.1 Longitudinal waves of collective order variables

Disregarding anomalies in the critical region, the collective order variable of pseudospins $\boldsymbol{\sigma}$ is represented by a classical vector, which is modulated at specific wavevector $\boldsymbol{q}$ and frequency $\tilde{\omega}$. Hence, we express the collective mode by the Fourier transform $\boldsymbol{\sigma}_q = \sigma_o e^{-i(\boldsymbol{q}\cdot\boldsymbol{r}-\tilde{\omega}\,t)}$, regarding crystal symmetry of the space group, for which an adiabatic potential $\Delta U_q = -\frac{A}{2}\boldsymbol{\sigma}_q^2 - \frac{B}{4}\sigma_o^4$ is responsible as inferred from Landau's theory. Therefore, a one-dimensional wave equation can be proposed to discuss the wave motion.

Following Krumshansl and Schrieffer [1], we write

$$m\left(\frac{\partial^2}{\partial t^2} - v_o^2\frac{\partial^2}{\partial x^2}\right)\sigma_q = \frac{\partial\,\Delta U_q}{\partial x} = -A\,\sigma_q + B\,\sigma_q^3 \tag{5.10c}$$

where $\tilde{\omega} = v_o q$, and m is the reduced mass of a displaced particle. Equation (5.10) was solved as follows.

Assuming $\sigma_q = \sigma_o e^{-i\phi}$ where $\phi = \boldsymbol{q}\cdot\boldsymbol{r} - \tilde{\omega}\,t$, equation (5.10) can be simplified by writing $\sigma_q/\sigma_o = Y$, i.e.

$$\frac{d^2Y}{d\phi^2} + Y - Y^3 = 0, \tag{5.10d}$$

where $\sigma_o = \sqrt{\frac{|A|}{B}}$, $\quad \boldsymbol{q}^2 = \frac{|A|}{m(v_o^2 - v^2)} = \frac{q_o^2}{1-(v/v_o)^2}$ and $q_o^2 = \frac{|A|}{mv_o^2}$.

Then we obtain from (5.10*b*) the *dispersion* relation

$$\tilde{\omega}^2 = v_o^2\left(\boldsymbol{q}^2 - \boldsymbol{q}_o^2\right). \tag{5.11}$$

Therefore, we have $\tilde{\omega} = 0$ for $\boldsymbol{q} = \boldsymbol{q}_o$. Signifying a phase transition, the phase for $T < T_c$ can be specified by $v < v_o$; for $T > T_c$ on the other hand, no collective motion is possible in the disordered phase. In the latter case, if the temperature T approaches close to T_c from above, the variable Y varies as a harmonic at a small amplitude.

Nevertheless, equation (5.10d) can be integrated as

$$2\left(\frac{\mathrm{d}Y}{\mathrm{d}\phi}\right)^2 = (\lambda^2 - Y^2)(\mu^2 - Y^2), \tag{5.12}$$

where $\lambda^2 = 1 - \sqrt{1-\alpha^2}$, $\mu^2 = 1 + \sqrt{1-\alpha^2}$ and $\alpha = \left(\frac{\mathrm{d}Y}{\mathrm{d}\phi}\right)_{\phi=0}$ is a constant.

Writing $\xi = \frac{Y}{\lambda}$ for convenience, equation (5.12) can be expressed by using an elliptic integral as

$$\frac{\phi_1}{\sqrt{2}\kappa} = \int_0^{\xi_1} \frac{\mathrm{d}\xi}{\sqrt{(1-\xi^2)(1-\kappa^2\xi^2)}}, \tag{5.13a}$$

where the upper limit ξ_1 corresponds to the phase ϕ_1. The constant $\kappa = \lambda/\mu$ is called the modulus, with which constants λ and μ are written as

$$\lambda = \frac{\sqrt{2}\,\kappa}{\sqrt{1+\kappa^2}} \quad \text{and} \quad \mu = \frac{\sqrt{2}}{\sqrt{1+\kappa^2}}.$$

The inverse function of (5.13) can then be written as

$$\xi_1 = \mathrm{sn}\frac{\phi_1}{\sqrt{2}\,\kappa}, \tag{5.13b}$$

which is a Jacobi's elliptic function for a longitudinal displacement σ_1 of the vector $\boldsymbol{\sigma}(\phi)$ described by a phase ϕ_1. Namely,

$$\sigma_1 = \lambda\sigma_0 \mathrm{sn}\frac{\phi_1}{\sqrt{2}\,\kappa}, \tag{5.13c}$$

where the amplitude is given by $\lambda\sigma_0$.

Next, considering the angular variable Θ_1 defined by $\xi = \sin\Theta$, the above relation (5.13b) allows for (5.13a) to change (5.13c) to a sinusoidal expression as $\mathrm{sn}\frac{\phi_1}{\sqrt{2}\,\kappa} = \sin\Theta_1$. In fact, Jacobi called $\Theta_1 = \mathrm{am}\left(\int_0^{\Theta_1}\frac{\mathrm{d}\Theta}{\sqrt{1-\kappa^2\sin^2\Theta}}\right)$ *an amplitude function*, however specifying *effective sinusoidal angle* for the nonlinear displacement $\sigma_1(\kappa)$. Figure 5.2 shows a graph of the relation between σ_1 and Θ for different values of κ.

The above theory predicted the presence of elliptic waves with finite amplitude in one-dimensional crystals of infinite length. However, we notice that a practical crystal includes surfaces, as cannot be infinite in size. Accordingly, Fourier's space must be applied to inside, whose surfaces are facilitated by Brillouin's conditions.

In fact, theoretically we can derive the expression

$$\sigma_1(\phi_1, \kappa = 1) = \sigma_0 \tanh\frac{\phi_1}{\sqrt{2}},$$

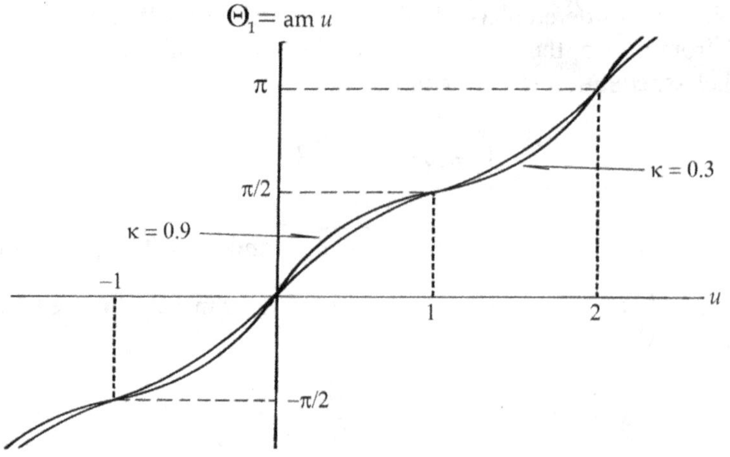

Figure 5.2. Jacobi's periodicity of a sn-function.

assuming to be the perpendicular component $\sigma_\perp(\phi_1)$ on surfaces, as permitted by the next section. For that, the theoretical assumption of one-dimensional theory should obviously be amended.

5.4.2 Transversal component and directional change of $\sigma(\phi)$

As experimentally confirmed, the order variable $\boldsymbol{\sigma}(\phi)$ is a classical vector. Therefore, both of the two components should be significant for nonlinearity of transition phenomena. At any rate, the transversal component can be written from the relation $\sigma_1^2 + \sigma_\perp^2 = \lambda^2\sigma_o^2$ as

$$\sigma_\perp = \lambda\sigma_o \cos\Theta_1 = \lambda\sigma_o \mathrm{cn}\frac{\phi_1}{\sqrt{2}\,\kappa} \quad \text{for } 0 < \kappa < 1 \tag{5.14}$$

confirming that (5.14) is the correct expression. Therefore, instead of $\sigma_1(\phi_1, \kappa = 1)$ in section 5.3.1, we have

$$\sigma_\perp(\phi_1, 1) = \pm\,\sigma_o \mathrm{sech}\frac{\phi_1}{\sqrt{2}} \quad \text{for} \quad \kappa = 1, \tag{5.15}$$

indicating the presence of singularity at T_c.

Now that the coherent pseudospin $\boldsymbol{\sigma}(\phi)$ is regarded as a classical vector, we have to verify their directional variation of *cnoidal potential energy* $V(\phi_1)$ along the longitudinal axis in crystals. At this point, it is noted that $V(\phi_1)$ is identical to the pinning potential $-V(\sigma_\perp)$ at singular site, hence we assume $\Delta V(\sigma_\perp) = \frac{A}{2}\sigma_\perp^2$ and no cross interaction between adjacent propagations. In contrast, for rotating $\sigma_\perp$ from $\phi_1 = -\Delta\phi$ to $\phi_1 = +\Delta\phi$, we need an energy

$$W = -\int_{-\Delta\phi}^{+\Delta\phi} \sigma_\perp \frac{\partial \Delta V(\sigma_\perp)}{\partial\phi_1}\mathrm{d}\phi_1 = \int_{-\Delta\phi}^{+\Delta\phi} \sigma_\perp \frac{\partial \Delta V(\phi_1)}{\partial\phi_1}\mathrm{d}\phi_1,$$

therefore, arriving at the following expression with (5.14), i.e.

$$\frac{\mathrm{d}W}{\mathrm{d}\phi_1} = -\frac{\lambda^2\sigma_0^2}{\kappa^2}\frac{\mathrm{d}}{\mathrm{d}\phi}\mathrm{dn}^2\frac{\phi_1}{\sqrt{2}\kappa} = \Delta V(\phi_1, \kappa).$$

Accordingly, the lattice potential can be expressed as

$$V(\phi_1, \kappa) = \frac{\lambda^2\sigma_0^2}{\kappa^2}\mathrm{dn}^2\frac{\phi_1}{\sqrt{2}\ \kappa}, \quad \text{for} \quad 0 < k < 1, \tag{5.16}$$

which is known as the *cnoidal potential* for the *cnoidal wave* (5.14).

The specific case $\kappa = 1$ expresses $V(\phi, 1) = \sigma_0^2 \mathrm{sech}^2\frac{\phi_1}{\sqrt{2}}$ from (5.16) however, accompanying with

$$V(\phi_1, 1) = \sigma_0^2 \tanh^2\frac{\phi_1}{\sqrt{2}} \quad \text{for } \sigma_0 \rightleftarrows -\sigma_0, \tag{5.17}$$

which is a solution of the Korteweg–deVries equation applied to practical crystals.

However, we notice that the mathematical process for $\kappa \to 1$ can be interpreted as the thermal process by time–temperature conversion. That is a significant postulate for $V(x, 1)$ to represent a thermodynamic potential.

5.4.3 Finite crystals and the domain structure

Since $\sigma_l(\phi_1)\perp\sigma_\perp(\phi_1)$, the phase $\frac{\phi_1}{\sqrt{2}}$ is a variable for changing from longitudinal $\sigma_l\left(\pm\frac{\pi}{2}\right)$ to transversal $\sigma_\perp\left(\mp\frac{\pi}{2}\right)$, signifying properties of *surfaces and the domain structure* in practical crystals. Accordingly, here we revise the redefined phase ϕ_1 for the range $-\frac{\pi}{2} < \phi_1 < \frac{\pi}{2}$ to relate with the inversion $\sigma_\perp \rightleftarrows -\sigma_\perp$, and the revised $\sigma_\perp$ and (5.17) are written as

$$\sigma_\perp(\phi_\perp, 1) = \sigma_0 \tanh\phi_\perp \quad \text{and} \quad V(\phi_\perp, 1) = \sigma_0^2\mathrm{sech}^2\phi_\perp \quad \text{at} \quad \kappa = 1. \tag{5.18}$$

At this point it is significant to notice that such a vector $\boldsymbol{\sigma}(T)$ with a finite magnitude occupies a special volume $\Delta\mathcal{V}$, creating a change of the internal energy $\Delta U = -p_{\mathrm{int}}\Delta\mathcal{V}$, where p_{int} is an effective internal pressure. Normally, a variation of $\Delta\boldsymbol{\sigma}$ at a given T can be studied as a thermodynamic transition for $\Delta n = 0$ at T, whereas for $\Delta n \neq 0$ is the transition is adiabatic. Accordingly, the latter should be investigated under variable external pressure p for $\Delta T = 0$. In this case, recalling the relation $\Delta U = -\mu\,\Delta n$, where μ is the chemical potential of soliton gas, the hypothetical p_{int} is regarded as related to Δn, and the relation $-p_{\mathrm{int}}\Delta\mathcal{V} = \mathcal{V}\Delta p$ allows to study $\Delta\boldsymbol{\sigma}$ as a function of T.

The above argument for surfaces and domains is mathematical but taking significant facts of finite crystals into account. In other words, the domain structure is regarded as a natural consequence of crystal surfaces, where the phase plays a basic role to specify the *generalized coordinates*. The following discussions will therefore be made for finite crystals with thermodynamic principles.

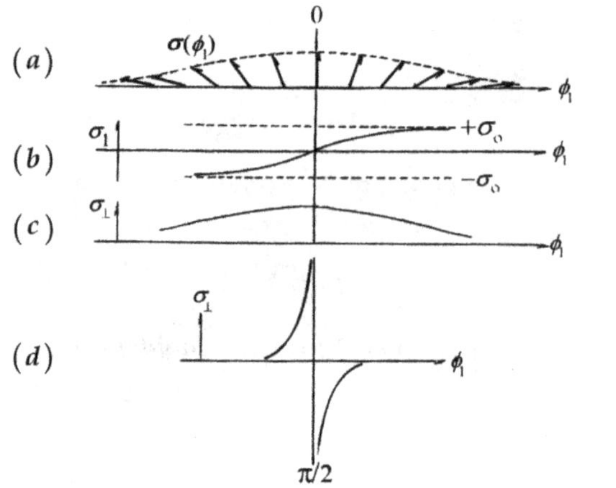

Figure 5.3. Partial views of nonlinear waves in uniform crystals. (*a*) Wave pattern of vectors. (*b*) Longitudinal component σ_1. (*c*) Transversal component $\sigma_\perp$. (*d*) Critical behavior of $\sigma_\perp$.

In figures 5.3(*a*)–(*c*), the curves of σ_1 and $\sigma_\perp$ are plotted near $\phi_1 = 0$ for $0 < \kappa < 1$ to illustrate their behaviors. Figure 5.3(*d*) shows $\sigma_\perp$ for $\kappa = 1$ near domain boundaries defined at $\phi_\perp = \pm\frac{\pi}{2}$.

For nonlinear dynamics, the complex expression $\sigma_1 + i\sigma_\perp$ is convenient, while dealing with entropy production for irreversible thermodynamics.

5.5 The complex lattice potential

In the field theory applied to crystals, the order variable is a moving object described by the density function

$$\boldsymbol{\sigma}(\phi_1) = \psi^*(\phi_1)\psi(\phi_1) \tag{5.19a}$$

of the object's wavefunction $\psi(\phi_1)$. In asymptotic approximation, we consider $\boldsymbol{\sigma}(\phi_1)$ is driven by the potential $V(\phi_1) = -K^2(\phi_1)\boldsymbol{\sigma}(\phi_1)$, where K is a function of ϕ_1. Similar to that assumed by P Weiss for a ferromagnetic crystal, this $V(\phi_1)$ is proportional to $\boldsymbol{\sigma}(\phi_1)$.

The motion is determined by the Klein–Gordon equation

$$\frac{\partial^2\boldsymbol{\sigma}}{\partial\tau^2} - v^2\frac{\partial^2\boldsymbol{\sigma}}{\partial x^2} = -K^2(x, \tau)\boldsymbol{\sigma}, \tag{5.19b}$$

where $\boldsymbol{\sigma}(x, \tau)$ is a complex variable to facilitate two-component equations (5.19*a*) and (5.19*b*), and v is the speed of wave propagation.

Now that $\boldsymbol{\sigma}(x, \tau)$ is complex, equation (5.19*b*) can be factorized as

$$\frac{\partial\psi}{\partial\tau} - v\frac{\partial\psi}{\partial x} = -iK(x, \tau)\psi^* \quad \text{and} \quad \frac{\partial\psi^*}{\partial\tau} + v\frac{\partial\psi^*}{\partial x} = -iK(x, \tau)\psi,$$

which are considered as development equations of $\psi = \psi' + i\psi''$ and $\psi^* = \psi'^* - i\,\psi''^*$. The development equations for ψ' and ψ'' can then be expressed as

$$\frac{\partial\psi'}{\partial\tau} - v\frac{\partial\psi'}{\partial x} = K(x,\tau)\psi'' \quad \text{and} \quad \frac{\partial\psi''}{\partial\tau} + v\frac{\partial\psi''}{\partial x} = K(x,\tau)\psi', \tag{5.20}$$

indicating a classical impact of the Klein–Gordon equation with a real potential.

Using Fourier's transformations

$$\psi' \pm i\psi'' = \{\Psi'(x,k) \pm i\Psi'(x,k)\}e^{\mp ikx} \text{ and } K(x,k) = K(x)e^{\mp ikx}$$

in crystal space, equation (5.20) can be re-expressed for the Fourier transforms (Ψ', Ψ'') as

$$\frac{d\Psi'}{dx} + ik\Psi' = u(x)\Psi'', \quad \frac{d\Psi''}{dx} - ik\Psi'' = -u(x)\Psi' \quad \text{and} \quad u(x) = -\frac{K(x)}{v}.$$

Further, writing complex variables $\Psi' \pm i\Psi'' = \psi_1$ and $\Psi' \mp i\Psi'' = i\psi_2$, we have

$$\frac{d\psi_1}{dx} \mp iu(x)\psi_1 = \mp k\psi_2 \quad \text{and} \quad \frac{d\psi_2}{dx} \pm iu(x)\psi_2 = \pm k\psi_1. \tag{5.21}$$

Eliminating ψ_2 and then ψ_1 from (5.21), we obtain

$$\frac{d^2\psi_1}{dx^2} + \left(k^2 \pm u^2 - i\frac{du}{dx}\right)\psi_1 = 0 \tag{5.22a}$$

and

$$\frac{d^2\psi_2}{dx^2} + \left(k^2 \pm u^2 + i\frac{du}{dx}\right)\psi_2 = 0, \tag{5.22b}$$

respectively.

By the above definitions, we can write $\psi_1^2 + \psi_2^2 = 0$. However, if writing $\sigma_1 = \psi_1^2$ and $\sigma_\perp = (i\psi_2)^2$, we have $\sigma_1^2 + \sigma_\perp^2 = 1$ for the vector $\boldsymbol{\sigma}$. Therefore, equations (5.22*a*) and (5.22*b*) can be revised as

$$\frac{d^2\sigma_1}{dx^2} + (k^2 \pm u^2)\sigma_1 = 0 \quad \text{and} \quad \frac{d^2\sigma_\perp}{dx} \mp \left(\frac{du}{dx}\right)\sigma_\perp = 0 \tag{5.23}$$

for a complex $\boldsymbol{\sigma} = \sigma_1 + i\,\sigma_\perp$ along the x-direction in crystals. Therefore, we have a complex energy $-u^2 \mp i\frac{du}{dx}$ for a complex $\boldsymbol{\sigma}$. Also, it is significant that the factor $K(x,\tau)$ is complex, although defined as real for the Weiss potential.

As suggested by (5.23) the presence of a *vertical imaginary potential* is proportional to $V_\perp(x) = \pm i\frac{du}{dx}$, as responsible for domain structure, which will be discussed later in the following chapters.

For thermodynamics, it is essential that the time conversion does not exist, while the space symmetry exists in nature. Nonlinear physics depends on symmetry significantly. Therefore, the *time* should now be converted to *temperature* to deal with the *nature.*

The *real potential* $V(\phi_c, T_c)$ defined in this section is a function of $x - v\,\tau$, where τ is converted to temperature T. Hence, the pinning potential $-V(\phi, T_c)$ is temperature dependent, and so is the lattice potential $+V(\phi, T_c)$, which signifies the temperature-dependent order variable $\boldsymbol{\sigma} = \sigma_l + i\,\sigma_\perp$. Accordingly, in the soliton theory, we can discuss the nature of $\boldsymbol{\sigma}$ by the equation

$$\frac{d^2\boldsymbol{\sigma}_{p,n}}{dz^2} + \{p^2 + n(n+1)\text{sech}^2 z\}\boldsymbol{\sigma}_{p,n} = 0, \quad (5.24a)$$

where $p = n, n-1, n-2, \ldots$.

Using the soliton number n, we write the equations for $\sigma_\perp(x, n)$ as

$$\frac{d^2\sigma_\perp(x, n)}{dx^2} + \{\lambda(n) + V(n)\}\sigma_\perp(x, n) = 0 \quad \text{for} \quad \sigma_\perp(x, n) = \sigma_\perp(x)e^{-i\tilde{\omega}\tau}, \quad (5.24b)$$

and boson operators $\tilde{b}_n(\tau) = \tilde{b}_n(0)e^{-i\tilde{\omega}\tau}$ and $\tilde{b}_n^\dagger(\tau) = \tilde{b}_n^\dagger e^{i\tilde{\omega}\tau}$, we express that

$$\sigma_\perp(x, n) = \tilde{b}_n(\tau)\sigma_n(x) \quad \text{and} \quad \sigma_\perp^*(x, n) = \tilde{b}_n^\dagger(\tau)\sigma_n^*(x),$$

from which the soliton Hamiltonian $\mathcal{H}_n$ can be calculated with the boson commutation relation as follows:

$$\left[\mathcal{H}_n, \tilde{b}_n^\dagger \tilde{b}_n\right] = 0 \quad \text{and} \quad \left[\tilde{b}_{n'}, \tilde{b}_n^\dagger\right] = \delta_{n',n},$$

showing that $\mathcal{H}_n = \lambda_n = n\lambda_o$, where λ_o is the lowest soliton energy. Therefore, the soliton potential energy can be expressed from (5.12) as

$$\pm V(\phi_1) = \pm \sigma_o^2 \text{sech}^2\phi_1 \quad (5.25)$$

and $-V(\phi_1)$ is for pinning energy at T_c.

5.6 Isothermal phase transition and entropy production

Equation (5.25) represents the result from a thermodynamic interpretation of Eckart's potential, which can logically be considered for the Weiss field substantiated by experiments. From this view, equation (5.25) represents a real potential supported by boson statistics.

Theoretically, however, the phase ϕ_1 is a fluctuating variable in a range between $-\Delta\phi_1$ and $+\Delta\phi_1$, so that the potential $V(\phi)$ represents the averaged value calculated over this range, i.e.

$$V(\phi) = \langle\mathcal{H}_n\rangle_{\text{thermal}} = n\lambda_o \propto \mp\langle\sigma_{\perp 1}\sigma_{\perp 2}\rangle_{\text{thermal}} = \left\langle\sigma_\perp^2\, \tilde{b}_{n1}^\dagger\tilde{b}_{n2}\right\rangle_{\text{thermal}}, \quad (5.26)$$

where the fluctuations in $\sigma_\perp$ andσ_l are illustrated in figure 5.4.

Exercises

1. It is significant that the order variable $\boldsymbol{\sigma}(\phi)$ is pinned by the Weiss potential $V(\phi)$ in phase in thermal equilibrium, as interpreted in terms of condensate pinning. Confirm from the derivation that $V(\phi)$ represents the Weiss field.

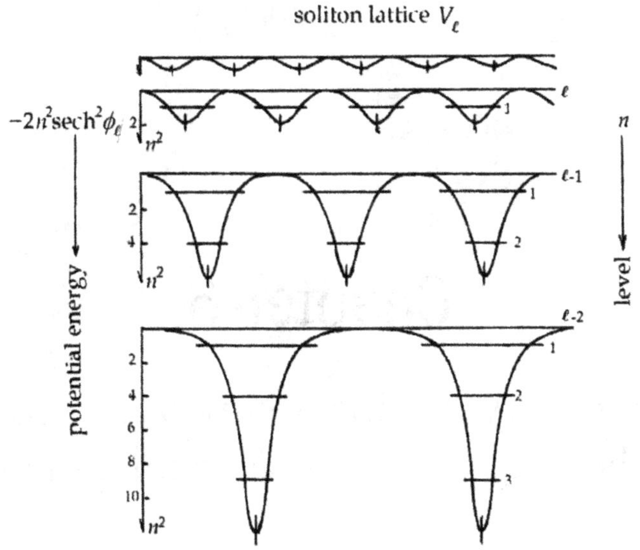

Figure 5.4. Spectral patterns of soliton lattices, showing association with soliton energies proportional to n^2, where n signifies soliton levels.

2. Recognized from (5.26) is that the negative potential $-V(\phi)$ should be characterized by the symmetric combination of $\sigma_{\perp 1}$ and $\sigma_{\perp 2}$ for thermal stability.
3. Discuss if we can consider non-Weiss internal fields among real systems?

Reference

[1] Krumshansl J A and Schrieffer J R 1975 *Phys. Rev.* B **11** 3535

IOP Publishing

Introduction to the Mathematical Physics of Nonlinear Waves
(Second Edition)

Minoru Fujimoto

Chapter 6

Soliton mobility in time–temperature conversion for thermal processes: Riccati's theorem

Soliton potentials have been detected experimentally as mobile solitary objects in crystals. Although limited to equilibrium and quasi-static phases, the mobility in Galilean invariance from a site to another space–time is measured in timescales of experiments, which is expressed for the wavevector in Brillouin zones. In addition however, it can be converted to a *temperature-scale* in thermodynamics.

Originating from lattice displacements, solitons are quantized objects as recognized by Riccati's transitions expressed by a few pulses between discrete levels in Eckart's potentials, behaving like free quasi-particles in crystals. On the other hand, solitons are characterized by *boson statistics* in a field-theoretical approximation.

In this chapter, the theoretical basis for soliton mobility in crystalline media is discussed with *time–temperature conversion*, which is necessary for thermodynamic lattice dynamics dealing with crystalline processes.

6.1 Bargmann's theorem

In this section, Bargmann's mathematical theorem in nonlinear dynamics from 1949 is discussed, showing that a steady Eckart's potential exhibits a wave in finite magnitude considering a soft-mode and related internal Weiss' potential in conservative dynamics in crystals. We discuss two Bargmann's theorems, following Lamb's textbook [1].

6.1.1 One-soliton solution

First, a wave equation $\frac{d^2\sigma}{dx^2} + q^2\sigma = 0$ for a free running wave $\sigma \sim e^{\pm iqx}$ as an elementary soliton at T, we consider a potential V for the lowest excitation at a wavevector q, and write the propagation equation of σ as

doi:10.1088/978-0-7503-3759-5ch6

$$\frac{d^2\sigma}{dx^2} + (q^2 - V)\sigma = 0 \tag{6.1}$$

where an amplitude-modulated wave is expressed as $\sigma = F(q, x)e^{iqx}$ under equilibrium conditions, with the amplitude $F(q, x)$ assumed as a polynomial of q. In this case, we write $\sigma = e^{iqx}\{2q + ia(x)\}$, where $V = 0$ at $q = 0$, and $a(x)$ is an increasing function of x; while $F(q, x)$ is increasing with q.

Differentiating σ for (6.1), we have $\frac{da}{dx} = -V$ and $\frac{d^2a}{dx^2} = Va$. Eliminating V from these relations, we obtain $\frac{d^2a}{dx^2} = -a\frac{da}{dx}$, hence $\frac{1}{2}\frac{d}{dx}\left(\frac{da}{dx}\right)^2 = -\frac{1}{2}\frac{d(a)^2}{dx}$. Integrating this, we have the relation $\frac{da}{dx} + \frac{a^2}{2} = 2\mu^2$, where μ^2 is a constant, if the system is conservative.

Transforming $a(x)$ by the relation $a = 2 \ln w$, the last equation can be expressed as linear, i.e.

$$\frac{d^2w}{dx^2} - \mu^2 w = 0$$

and its solution is given by $w = \alpha\, e^{\mu x} + \beta\, e^{-\mu x}$, where α and β are arbitrary. In this case, from (6.1) we have

$$V = -2\frac{d^2(\ln w)}{dx^2} = -2\mu^2\text{sech}^2(\mu x - \theta) \quad \text{where } \theta = \frac{1}{2}\ln\frac{\beta}{\alpha},$$

which is Eckart's potential obtained as a steady-state solution of Korteweg–deVries' equation at $\kappa = 1$ for a given temperature. Substituting this into (6.1), we see that $\theta = 4\mu^3 t$, and the Eckart potential can be expressed as

$$V = -2\mu^2\text{sech}^2(\mu x - 4\mu^3 t), \tag{6.2}$$

agreeing with the corresponding formula, if we set $\mu = \frac{\sqrt{v}}{2}$, since the phase of (6.2) is $\phi = \mu x - \theta = \frac{\sqrt{v}}{2}(x - vt)$ under an *isothermal transformation* $x - vt \to x'$ at $\Delta T = 0$ and the speed v is constant, while $\mu \neq \frac{\sqrt{v}}{2}$ corresponds to the *non-conservative adiabatic transition* $\Delta T \neq 0$.

The Bargmann potential (6.2) at a given temperature T carries an important soliton feature of *coherent wave* as signified by a finite amplitude specified by the number of solitons. Also, representing collective motion of order variables, space invariance signified by discrete changes in $\mathbf{q} \to \mathbf{q} + \mathbf{G}$ indicate a Galilean mobility in momentum space, characterized by fluctuating $\Delta\mathbf{q}$. Mathematically, (6.2) is only a specific solution for $\kappa = 1$, however, it has been confirmed physically to be involved with many thermodynamic processes.

6.1.2 Two-soliton solutions

Assuming that Bargmann's amplitude function $F(q, x)$ is expressed in the second order of q, we can also show that the potential V satisfying the Korteweg–deVries equation can be composed of two independent Eckart's potentials, just by the superposition principle, to consider an adiabatic soliton gas. Here, ignoring all possible damping effects for simplicity, we assume that

$$\sigma = e^{iqx}\left\{4q^2 + 2ia(x) + b(x)\right\}, \tag{6.3}$$

which is substituted in (6.1), and we obtain

$$-\frac{\mathrm{d}a}{\mathrm{d}x} = V, \quad \frac{\mathrm{d}^2a}{\mathrm{d}x^2} + \frac{\mathrm{d}b}{\mathrm{d}x} = Va \quad \text{and} \quad \frac{\mathrm{d}^2b}{\mathrm{d}x^2} = Vb,$$

which can be manipulated for these expressions to be integrated. Eliminating V from above, we obtain a relation $\frac{\mathrm{d}}{\mathrm{d}x}\left(-\frac{a^2}{2} - \frac{\mathrm{d}a}{\mathrm{d}x}\right) = \frac{\mathrm{d}b}{\mathrm{d}x}$, which is integrated as $b + \frac{\mathrm{d}a}{\mathrm{d}x} + \frac{a^2}{2} = 2c_1$ where c_1 is a constant. Also, another relation $b\left(\frac{\mathrm{d}^2a}{\mathrm{d}x^2} + \frac{\mathrm{d}b}{\mathrm{d}x}\right) - a\frac{\mathrm{d}^2b}{\mathrm{d}x^2} = 0$ is integrated as $\frac{b^2}{2} + b\frac{\mathrm{d}a}{\mathrm{d}x} - a\frac{\mathrm{d}b}{\mathrm{d}x} = 2c_2$ where c_2 is a constant.

Writing $a = 2\frac{\mathrm{d}\ln w}{\mathrm{d}x}$ as before, from the first integrated result we derive $b = 2\left(c_1 - \frac{1}{w}\frac{\mathrm{d}^2w}{\mathrm{d}x^2}\right)$, and from the second relation we obtain

$$2\frac{\mathrm{d}w}{\mathrm{d}x}\frac{\mathrm{d}^3w}{\mathrm{d}x^3} - \left(\frac{\mathrm{d}^2w}{\mathrm{d}x^2}\right)^2 - 2c_1\left(\frac{\mathrm{d}w}{\mathrm{d}x}\right)^2 + w^2\left(c_1^2 - c_2\right) = 0.$$

The last equation can then be re-expressed as

$$\frac{\mathrm{d}^4w}{\mathrm{d}x^4} - 2c_1\frac{\mathrm{d}^2w}{\mathrm{d}x^2} + \left(c_1^2 - c_2\right)w = 0, \tag{6.4}$$

which can be solved in the form $w \sim e^{\Omega x}$ where $\Omega^2 - c_1 \pm \sqrt{c_2}$. We therefore express the four roots of the integrated equation by $\Omega_1 = \pm\sqrt{c_1 + \sqrt{c_2}}$ and $\Omega_2 = \sqrt{c_1 - \sqrt{c_2}}$, and (6.4) has a solution

$$w = (\alpha_1 e^{\Omega_1 x} + \beta_1 e^{-\Omega_1 x}) + (\alpha_2 e^{\Omega_2 x} + \beta_2 e^{-\Omega_2 x}).$$

Here, these four constants α_1, β_1, α_2 and β_2 are not independent, as related by $\alpha_1\beta_1\Omega_1^2 = \alpha_2\beta_2\Omega_2^2$, therefore, the above equation can be expressed by

$$w = 2\,\Omega_2\cosh(\Omega_1 x - \varphi_1) + 2\Omega_1\cosh(\Omega_2 x - \varphi_2).$$

Accordingly, $a = 2\frac{\mathrm{d}\ln w}{\mathrm{d}x} = 2\,\Omega_1\Omega_2\frac{\sinh(\Omega_1 x - \varphi_1) + \sinh(\Omega_2 x - \varphi_2)}{\Omega_2\cosh(\Omega_1 x - \varphi_1) + \Omega_1\cosh(\Omega_2 x - \varphi_2)}$.

Further, writing that $\Omega_1 = q_1 + q_2$, $\Omega_2 = q_1 - q_2$, $\varphi_1 = \theta + \chi$ and $\varphi_2 = \theta - \chi$, we obtain $a = 2\left(q_1^2 - q_2^2\right)\left(q_1\coth\phi_1 - q_2\tanh\phi_2\right)^{-1}$ where $\phi_1 = q_1 x - \theta$ and $\phi_2 = q_2 x - \chi$.

Hence $V = -\frac{\mathrm{d}a}{\mathrm{d}x} = -2\left(q_1^2 - q_2^2\right)\frac{q_1^2\mathrm{cosech}^2\phi_1 + q_2^2\mathrm{sech}^2\phi_2}{(q_1\coth\phi_1 - q_2\tanh\phi_2)^2}$.

Assuming that $\phi_1 \sim 1$ and $\phi_2 \sim 0$ in the above, we obtain

$$V \approx V_\perp(q_1) + V_\perp(q_2) \tag{6.5a}$$

where

$$V_\perp(q_1) = -2q_1^{\,2}\mathrm{sech}^2(q_1 x - \theta \mp \Delta),\ \ V_\perp(q_2) = -2q_2^{\,2}\mathrm{sech}^2(q_2 x - \chi \mp \Delta) \tag{6.5b}$$

and $\Delta = \tanh \frac{q_2}{q_1}$; the phase parameter θ and χ can be functions of fluctuation time τ, as is consistent with the one-soliton solution.

Equation (6.5) represents *two soliton pulses* that can interact each other, while separated as signified by different wavevectors. However, when colliding their wavevectors become $q_1 = q_2$, exhibiting coherent characters of solitons, as illustrated in figure 6.1, where the phase relations are also demonstrated by computation.

At such a crossing point, if the pulse intensity can be sufficiently high, the soliton intensity may become critical energy transfer to the surroundings, as described by entropy production in the system.

6.2 Riccati's theorem and the modified Korteweg–deVries equation

6.2.1 Riccati's theorem

Nonlinear waves are pinned by a soliton potential in phase, as shown by the Korteweg–deVries equation. On the other hand, for the soliton potentials the

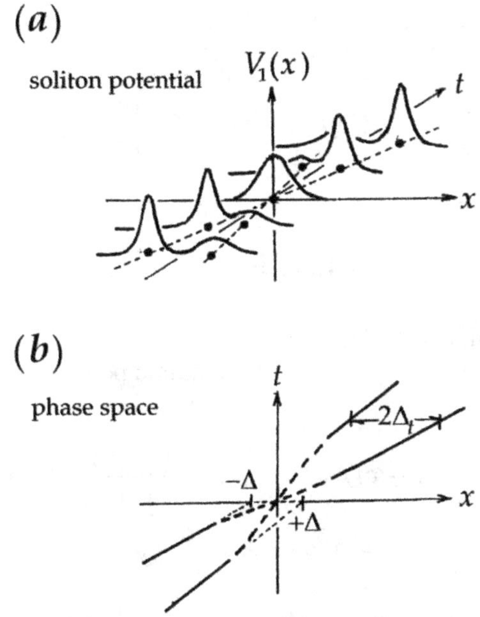

Figure 6.1. Bargmann's two-soliton solutions obtained numerical calculations. Reproduced from [1] John Wiley & Sons. Copyright © 1981 WILEY-VCH Verlag GmbH & Co. KGaA, Weinheim.

Riccati theorem signifies the mobile pathway of pinned energy from a soliton pulse to another

Considering two values $V_1(x)$ and $V_2(x)$ of the potential $V(x)$, the corresponding waves of order variables $\sigma_1(x)$ and $\sigma_2(x)$ are connected linearly as

$$\sigma_1 = A(x, \lambda)\sigma_1 + B\frac{d\sigma_2}{dx} \tag{6.6a}$$

where B is assumed to be a small constant and

$$\frac{d^2\sigma_1(x)}{dx^2} = \{-\beta^2 + V_1(x)\}\sigma_1(x) \quad \text{and} \quad \frac{d^2\sigma_2(x)}{dx^2} = \{-\beta^2 + V_2(x)\}\sigma_2(x). \tag{6.6b}$$

Substituting (6.6a) into (6.6b), we obtain

$$A_{xx} + V_{1x} + A(V_1 - V_2) = 0 \quad \text{and} \quad 2A_x + (V_1 - V_2) = 0. \tag{6.6c}$$

Eliminating $V_1 - V_2$ from (6.6c), we derive $A_{xx} - 2A_xA - V_{1x} = 0$, which is integrated as

$$A^2 - A_x - V_1 = -\bar{\beta}^2 \tag{6.6d}$$

where $\bar{\beta}^2$ is a constant of integration. Known as *Riccati's equation*, this expression is linearized by transforming A to $\tilde{\sigma}$ defined by $A = -\frac{d \ln \tilde{\sigma}}{dx}$, resulting in

$$\frac{d^2\tilde{\sigma}(x)}{dx^2} = \{-\bar{\beta}^2 + V_1(x)\}\tilde{\sigma}(x)$$

which is identical to the first equation in (6.6b), provided that $\bar{\beta} = \beta$. In this case, the second equation in (6.6b) can be written as

$$\frac{d^2\sigma_2(x)}{dx^2} = \left\{-\beta^2 + V_1(x) - 2\frac{d^2(\ln \tilde{\sigma})}{dx^2}\right\}\sigma_2(x),$$

leading to the transition

$$V_2(x) - V_1(x) = -2\frac{d^2(\ln \tilde{\sigma})}{dx^2}, \tag{6.6e}$$

which represents significant *soliton mobility*.

Considering (6.6e) as changes in thermodynamic potentials, we need to deal with the differential

$$\begin{aligned} \Delta V(x - vt) &= \left(\frac{\partial V}{\partial x}\right)_t \Delta x - v\left(\frac{\partial V}{\partial t}\right)_x \Delta t \\ &= 2\frac{\tilde{\sigma}'}{\tilde{\sigma}}\Delta x - \frac{2v\,\tilde{\sigma}'}{\tilde{\sigma}}\Delta t, \end{aligned} \tag{6.7}$$

where the second term for Δt represents an *isothermal change* with *time–temperature conversion* $\Delta t \propto \Delta T$, assuming $\tilde{\sigma}'/\tilde{\sigma}$ is constant. On the other hand, the first term for

Δx signifies an *adiabatic change*, as exemplified by $n \to n - 1$ under this assumption. The adiabatic transition is normally signified by a change in soliton numbers Δn. Figure 6.1 shows a typical thermodynamic change in observed heat capacity below critical temperature T_c.

6.2.2 Modified Korteweg–deVries equations in a conservative system

In this section, we emphasize the fact that the two-component field in a conservative system expressed by $\sigma = \sigma' + i\sigma''$ is signified by the complex potential energy $V(x) = u^2 \pm i\frac{du}{dx}$, where $u(x) = -\frac{K(x)}{v}$. On the other hand, the development equation is not necessarily compatible with the Korteweg–deVries equation, although $V(x)$ should be in free propagation. Therefore, as required by the relation $\frac{\partial V}{\partial t} - v\frac{\partial V}{\partial x} = 0$, we have derived Korteweg–deVries' equation $\frac{\partial V}{\partial t} - 6V\frac{\partial V}{\partial x} + \frac{\partial^3 V}{\partial x^3} = 0$ specified by $v = 6V$.

On the other hand, if combining the above relations mathematically, we can obtain a new expression

$$\frac{\partial u}{\partial t} - 6u^2\frac{\partial u}{\partial x} + \frac{\partial^3 u}{\partial x^3} = 0, \tag{6.8a}$$

which is called *the modified Korteweg–deVries equation*, nevertheless converted by transformation of an imaginary potential $u = -i\bar{u}$ as

$$\frac{\partial \bar{u}}{\partial t} + 6\bar{u}^2\frac{\partial \bar{u}}{\partial x} + \frac{\partial^3 \bar{u}}{\partial x^3} = 0. \tag{6.8b}$$

Using Riccati's transformation $\bar{u} = -\frac{d \ln \sigma}{dx}$ in (6.8*b*), we can define a complex potential $\bar{V} = V_\parallel + iV_\perp = -\bar{u}^2 - i\frac{d\bar{u}}{dx}$ in (6.8*b*), we derive steady-state equations

$$\frac{d^2\sigma}{dx} - \bar{V}\sigma = 0 \quad \text{and} \quad \frac{d^2\sigma}{dx^2} - (\bar{V} + \Delta\bar{V})\sigma = 0,$$

which are identical for a constant $\Delta\bar{V}$, constituting *Galilean invariance* at constant T. At this point, it is significant that the Galilean invariance corresponds to thermodynamic identity by conversion of the potential $\bar{V}(x)$ at T.

6.3 Soliton mobility studied by computational analysis

The significance of the soliton potential energy behaving like pulses of (6.2) and (6.5*b*) expressed by x and t was first recognized by computational studies of Zabusky and Kruskal [2]. They studied the variables a and b of Bargmann's theory numerically and established solitons as mobile objects in pulses in *conservative crystalline media*.

In deriving Korteweg–deVries' equation, they selected $a = -4$ in mathematical convenience, resulting in the term in the third-order derivative $\frac{\partial^3 V}{\partial x^3}$ that is responsible for dispersive propagation. To deal with the initial situation, the frequency was specified as $\omega = v_0 k - \mu k^3$ for the initial phase to specify by v_0 and dispersion parameter μ. Accordingly, the phase velocity at time t was considered as determined by

$$\begin{aligned} v &= v_0 + \mu^2 v_1 = v_0 + 2\mu^2 \mathrm{sech}^2\{\mu(x - v_0 t) - 4\mu^3 t\} \\ &= v_0 + 2\mu^2 \mathrm{sech}^2(\mu x - \omega t), \end{aligned} \tag{6.9a}$$

where the phase in { … } is discontinuous, since $v = v_0$ and $\mu = 0$ at $t = t_0$.

On the other hand, it was considered that (6.9*a*) represents sinusoidal peaks of the lattice expressed by

$$v = v_0 + 2\mu^2 \cos(\mu x - \omega t), \tag{6.9b}$$

because nodal point solitons in both (6.9*a*) and (6.9*b*) should coincide on the x-axis.

In their numerical analysis, Zabusky and Kruskal considered that a sinusoidal wave

$$v = \cos \pi(x - t) \quad \text{for} \quad t < t_0 \tag{6.9c}$$

determines the nodal points before the critical time t_0 at $x = 0$ and $\frac{\pi}{2}$ at $t_B = \frac{1}{\pi} < t_0$.

Figure 6.2 shows numerical results for a very weak dispersion $\sqrt{\mu} = 0.022$, where pulse amplitudes are plotted at $t = 0$, t_B and $3.6t_B$, where t_B is known as the development timescale. At these times, the emergence of soliton peaks is clearly seen as a simulated pattern, where they identified eight solitons at $3.6t_B$.

Significant in this diagram is that those soliton potentials are mobile in the phase (x, t), indicating the mobility of sech^2–pulses in the dynamical phase space. These soliton peaks overlap at crossing points with no change in shapes and amalgamated at $t = 0.5\ t_B$, showing a small number of peaks in larger amplitudes. Illustrated in figure 6.3, such collision of two solitons implies *coherence* due to nonlinearity. We therefore interpret that (6.9*c*) represents the static equilibrium of correlation energy in the system, where the soliton motion is described like free particles. We may

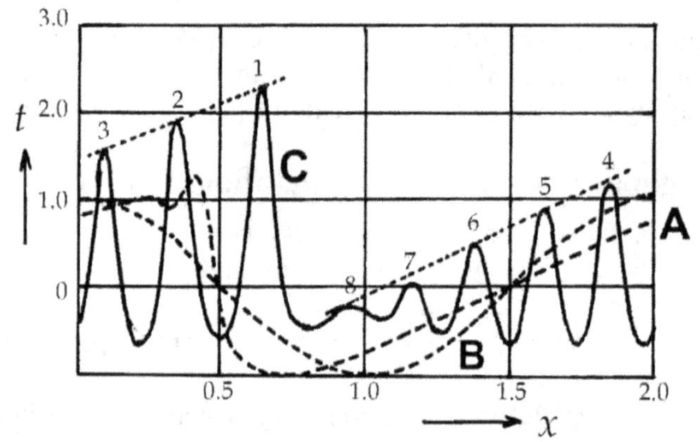

Figure 6.2. Eight soliton peaks at different times indicated by 1, 2, 3, …, 7, 8. Reprinted with permission from [2]. Copyright (1965) by the American Physical Society.

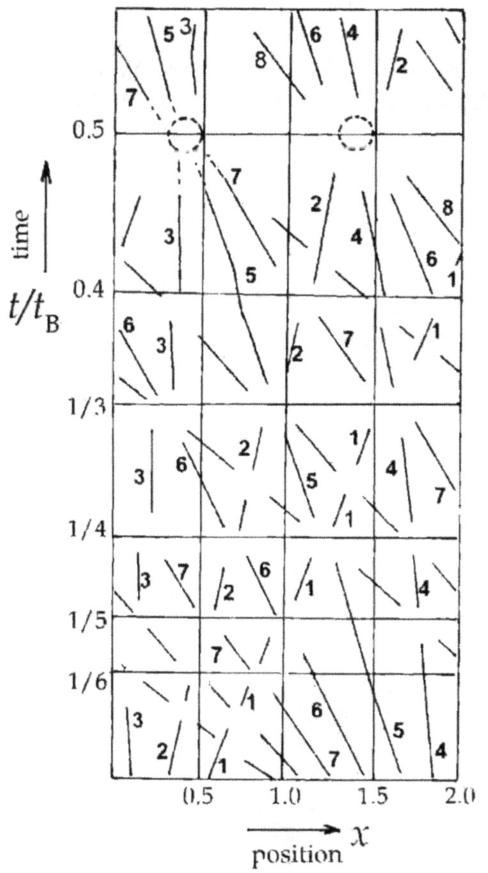

Figure 6.3. Loci of soliton peaks numerically calculated in the dynamical phase diagram. Reprinted with permission from [2]. Copyright (1965) by the American Physical Society.

therefore assume that the specific time 0.5 t_B represents a critical temperature for structural change of *soliton lattice*.

It is essential to realize that these computational studies showed that the space–time mobility of soliton energies in boson statistics in arbitrary directions of crystals is coherent as restricted to symmetric combinations, providing stress as well as temperature distributions in the soliton system.

Exercises

1. Why are solitons considered as mobile particles in conservative systems? Discuss their mobility with respect to physical principles.
2. Discuss the differences between Korteweg–deVries' equation and its modified form. Show the mathematical and physical differences.
3. With respect to the phase function ϕ in soliton theory, it is important to deal with time–temperature conversion for the thermodynamic argument. Discuss the importance of isothermal and adiabatic transitions.

References

[1] Lamb G L Jr 1980 *Elements of Soliton Theory* (New York: Wiley)
[2] Zabusky N J and Kruskal M D 1965 *Phys. Rev. Lett.* **15** 240

IOP Publishing

Introduction to the Mathematical Physics of Nonlinear Waves
(Second Edition)

Minoru Fujimoto

Chapter 7

Toda's lattice of correlation potentials

In previous chapters, a vector model of variable order with a finite amplitude of displacement was discussed for modulated crystals. Although substantiated experimentally, such a model is inadequate for analyzing nonlinearity in finite crystals, because existing anomalies would be subjected to *physical uncertainties*. Toda [1] discovered an *exponential potential* in the correlation energy to remove such a mathematical problem, so scatterings from Eckart's potentials can be analyzed algebraically. In this chapter, we introduce Toda's potential, thus the development of nonlinearity can be analyzed algebraically.

7.1 The Toda soliton lattice

7.1.1 Dual chains of condensates

In chapter 3, we discussed a long chain of identical displacements in a one-dimensional structure. Such displacements can take place against counteracting ions at lattice sites, moving together in phase, which is known as a *condensate*. Characterized by the reduced mass, pseudospins and the counteracting lattice are in collective motion, which are described as the *dual*.

Figure 7.1 shows the model dual chains, where one chain is specified by mass m and spring constant κ_A, and another chain similarly by M and κ_B. Writing their displacements from lattice site n by r_n and R_n, we have the momentum conservation law $m\dot{r}_n + M\dot{R}_n = 0$.

For the dual of chains, A and B, a Hamiltonian can be expressed for chain A as

$$\mathcal{H}_A = \frac{1}{m}\sum_n p_n^2 + \sum_n \phi(r_n), \tag{7.1a}$$

doi:10.1088/978-0-7503-3759-5ch7

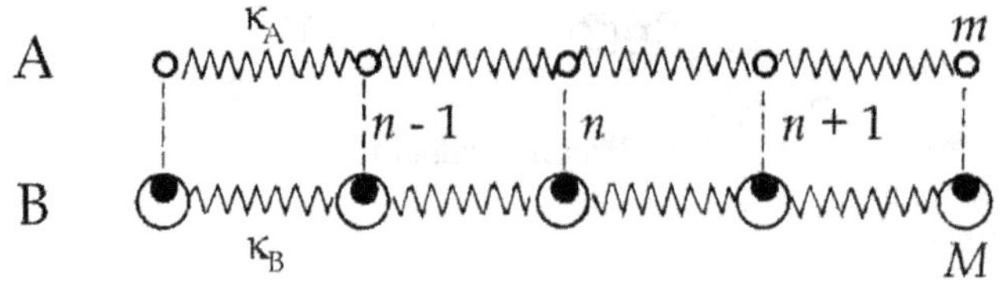

Figure 7.1. Toda's dual structure, representing a condensate pair for binary transitions. Shown is a dual A of mass m and another dual B of mass M with the same spring constant κ, where interactions occur at $n-1$, n, $n+1$ sites.

where $p_n = \frac{\partial K_A}{\partial y_n} = m\dot{y}_n$ and $K_A = \frac{1}{2}\sum_n m\dot{y}_n^2$ is the kinetic energy, taking y as direction for vibration; hence $y = y_0$ is set on the x-axis.

Assuming $y_0 = 0$ for simplicity, we write

$$y_0 = 0,\ y_1 = r_0,\ y_2 = r_0 + r_1,\ \ldots;$$
$$\dot{y}_0 = 0,\ \dot{y}_1 = \dot{r}_0,\ \dot{y}_2 = \dot{r}_0 + \dot{r}_1,\ \ldots.$$

For sites $n = 0, 1, \ldots, N-1$ of the chain A, the kinetic energy of collective motion is

$$K_A = \frac{m}{2}\sum_{n=0}^{N-1}(\dot{r}_0 + \dot{r}_1 + \cdots + \dot{r}_n)^2,$$

whose conjugate momentum can be defined as

$$s_n = \frac{\partial K_A}{\partial r_n} = m\sum_{n=0}^{N-1}(\dot{r}_0 + \dot{r}_1 + \cdots) = m\dot{y}_n,$$

hence

$$s_{n-1} - s_n = m\dot{y}_n \quad \text{and} \quad s_N = 0.$$

The Hamiltonian can then be expressed as

$$\mathcal{H}_A = \frac{1}{2m}\sum_{n=0}^{N-1}(s_{n+1} - s_n)^2 + \sum_{n=0}^{N-1}\phi(r_n),$$

for which the canonical equations of motion are written as

$$\dot{r}_n = \frac{\partial \mathcal{H}_A}{\partial s_n} = -\frac{s_{n+1} - 2s_n + s_{n-1}}{m} \quad \text{and} \quad \dot{s}_n = -\frac{\partial \mathcal{H}_A}{\partial r_n} = -\phi'(r_n). \tag{7.2}$$

Eliminating s_n from these equations, we obtain

$$m\ddot{r}_n = \phi'(r_{n+1}) - 2\phi'(r_n) + \phi'(r_{n-1}), \tag{7.3}$$

describing the collective motion in the lattice A.

If the potential $\phi(r_n)$ is harmonic, it is characterized by the spring constant κ_A, so that (7.3) is expressed as $m\ddot{r}_n = \kappa_A(r_{n+1} - 2r_n + r_{n+1})$, while for the counteracting lattice B, (7.3) is written similarly as $M\ddot{R}_n = \kappa_B(R_{n+1} - 2R_n + R_{n-1})$, constituting the dual.

On the other hand, (7.2) can be used for restoring potential in the other type, for which the second equation is modified by

$$mr_n = -\chi(\dot{s}_n), \tag{7.4a}$$

deriving the equation

$$\frac{d}{dt}\chi(\dot{s}_n) = s_{n+1} - 2s_n + s_{n-1}. \tag{7.4b}$$

Using the law of momentum conservation, (7.4*a*) can be modified to the lattice B as $MR_n = \chi(\dot{S}_n)$ where $S_n = \langle s_n \rangle_t$, taking possible fluctuations into account.

Corresponding to (7.1) for lattice A, we can write the Hamiltonian

$$\mathcal{H}_B = \frac{1}{2M}\sum_n P_n^2 + \sum_n \phi(R_n), \tag{7.1b}$$

for which the coordinates are expressed as

$$Y_n = \frac{1}{M}(S_{n-1} - S_n) \quad \text{and} \quad R_n = -\frac{1}{M}(S_{n+1} - 2S_n + S_{n-1}),$$

and

$$\chi(\dot{S}_n) = S_{n+1} - 2\, S_n + \, S_{n-1}. \tag{7.4c}$$

These s_n and S_n in (7.4*b*) and (7.4*c*) represents displacements in a dual of A and B, respectively, hoping to be responsible for the physical property of order variable $\boldsymbol{\sigma}_n$ of a condensate.

7.1.2 Toda's correlation potentials

It is now considered that the dual should be placed relative to the Eckart potential, for which the function χ in (7.4*c*) is required to be found. Toda [1] discovered an exponential form of the function ϕ to satisfy (7.4*c*). The procedure for Toda's potential ϕ was mathematically tedious but is outlined as follows.

Considering the fact that Eckart' potential originated from the soliton potential $\text{sn}^2\, u$, he first used the formula known as the addition theorem for order variables u and v,

i.e.

$$\text{sn}^2\,(u + v) - \text{sn}^2\,(u - v) = 2\frac{d}{dv}\frac{\text{sn}\, u\; \text{cn}\, u\; \text{dn}\, u\; \text{sn}^2\, v}{1 - \kappa^2 \text{sn}^2\, u\; \text{sn}^2\, v}.$$

Here, using $\text{dn}^2\, u = 1 - \kappa^2\text{sn}^2\, u$, the function $\varepsilon(u) = \int_0^u \text{dn}^2\, u\; du$ is defined.

Using the relations

$$\varepsilon'(u) = \text{dn}^2\, u \quad \text{and} \quad \varepsilon''(u) = -2\,\kappa^2 \text{sn}\, u\; \text{cn}\, u\; \text{dn}\, u,$$

we obtain

$$\varepsilon(u+v) - 2\varepsilon(u) + \varepsilon(u-v) = \frac{\varepsilon''(u)}{\frac{1}{\mathrm{sn}^2 v} - 1 + \varepsilon'(u)}. \tag{7.5}$$

While $\varepsilon(u)$ is not a periodic function, Jacobi's zeta-function defined by $Z(u) = \varepsilon(u) - \frac{E}{K}$ is periodic with period $2K$, where K and E are complete elliptic integrals of the 1st and 2nd kinds, respectively. Rewriting (7.5) with $Z(u)$, we have

$$Z(u+v) - 2Z(u) + Z(u-v) = \frac{\mathrm{d}}{\mathrm{d}u} \ln\left(1 + \frac{Z'(u)}{\frac{1}{\mathrm{sn}^2 v} - 1 + \frac{E}{K}}\right). \tag{7.6a}$$

Now we can see that (7.6) can be compared with (7.4*a*) and (7.4*b*), if we set

$$u = 2\left(\tilde{\nu}\tau \pm \frac{n}{\tilde{\lambda}}\right)K, \quad v = \frac{2K}{\tilde{\lambda}}, \quad s_n(\tau) = \frac{2K\tilde{\nu}}{b/m} Z(u), \tag{7.6b}$$

where b is a constant, and we obtain for s_n as follow, representing the representative variable s. Namely,

$$\chi(\dot{s}) = \frac{m}{b} \ln\left(1 + \frac{\frac{b/m}{(2K\tilde{\nu})^2}}{\frac{1}{\mathrm{sn}^2 v} - 1 + \frac{E}{K}} \dot{s}\right) - ma, \tag{7.7}$$

where a is another constant, and by setting the expression

$$(2K\tilde{\nu})^2 = \frac{ab}{m}\left\{\frac{1}{\mathrm{sn}^2(2K/\tilde{\lambda})} - 1 + \frac{E}{K}\right\}^{-1}$$

we obtain $r = -\frac{1}{b}\ln\left(1 + \frac{\dot{s}}{a}\right) + r_0$, where r_0 may be regarded as the parameter for nonlinearity.

Therefore, from (7.7) we have

$$\dot{s} = a(e^{-b(r-r_0)} - 1) = -\phi'(r). \tag{7.8a}$$

Accordingly, Toda's potential is expressed as

$$\phi(r) = \frac{a}{b}e^{-b(r-r_0)} + ar + \text{const. or } \phi(r) = Ae^{-br} + ar, \quad \text{where } A = \frac{a}{b}e^{br_0}. \tag{7.8b}$$

7.1.3 Propagation in Toda's soliton lattice

It is practical to consider $r_0 = 0$ for the potential $\phi(r)$ in (7.8) to discuss the soliton lattice in crystals. Therefore, in this section we express Toda's potential as

$$\phi(r) = \frac{a}{b}e^{-br} + ar, \quad \text{where} \quad ab > 0. \tag{7.8b}$$

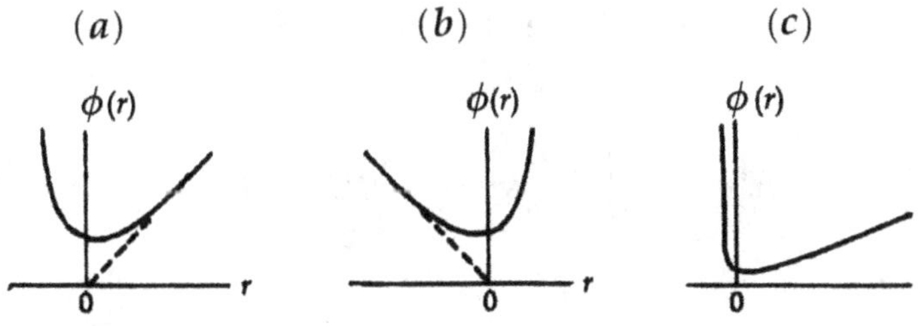

Figure 7.2. Toda's correlation potential expressed as $\phi = \frac{a}{b}e^{-br} + ar$: (*a*) for *a*, $b > 0$, (*b*) for *a*, $b < 0$ and (*c*) for a large *b*.

Figure 7.2 shows the property of $\phi(r)$ for $ab > 0$, consisting of a repulsive force and a nonlinear feature, near $r = 0$ and a distant place, respectively. Figure 7.2(*c*) is for a large *b*, and figures 7.2(*a*) and 7.2(*b*) are for $+r$ and $-r$, so basically the same. Such additive parts in (7.8*b*) can be studied separately, so $\phi(r)$ is in a convenient form experimentally.

Another advantage is that Toda's potential can be applied to cases under external pressure. With an external force *f*, we may consider $a' = a + f$ as an effective force, and $a = a'e^{-ba'}$, so that (7.8*b*) can be written as

$$\phi(r) = \frac{a'}{b}e^{-b(r+a')} + a'\,r, \tag{7.8c}$$

which is in the same form as (7.8*b*).

In any case of dealing with collective motion, we notice that the second terms in (7.8*b*) and (7.8*c*) do not appear in equations (7.1*a*) and (7.1*b*), respectively. Therefore, the equation of motion for the lattice A can be written as

$$m\frac{\mathrm{d}^2 y_n}{\mathrm{d}\tau^2} = a\left(e^{-b(y_n - y_{n-1})} - e^{-b(y_{n+1} - y_n)}\right)$$

and

$$m\frac{\mathrm{d}^2 r_n}{\mathrm{d}\tau^2} = a\left(2e^{-br_n} - e^{-br_{n-1}} - e^{-br_{n+1}}\right).$$

For the dual, (7.4*b*) is expressed as

$$\frac{\mathrm{d}}{\mathrm{d}\tau}\ln(a + \dot{s}_n) = \frac{\ddot{s}_n}{a + \dot{s}_n} = \frac{b}{m}(s_{n-1} - 2s_n + s_{n+1}) \tag{7.9a}$$

and from (7.2) $\langle \dot{s}_n \rangle = f_n$ satisfies the expression

$$\frac{\mathrm{d}^2}{\mathrm{d}\tau^2}\ln\left(1 + \frac{f_n}{a}\right) = \frac{b}{m}(f_{n-1} - 2f_n + f_{n+1}), \tag{7.9b}$$

describing the *nonlinear spring force* in the dual (A, B). Therefore, for the lattice B, we have

$$\ln\left(1 + \frac{\dot{S}_n}{a}\right) = \frac{b}{m}(S_{n-1} - 2S_n + S_{n+1}), \tag{7.9c}$$

and

$$f_n = a(e^{-br_0} - 1) = \langle \dot{s}_n \rangle. \tag{7.9d}$$

The periodicity of $s_n(\tau)$ for nonlinear waves is determined by the zeta function $Z(u)$ defined in (7.6*b*). Therefore, we write

$$e^{-br_n} - 1 = \frac{(2K\tilde{\nu})^2}{ab/m}\left[\mathrm{dn}^2\left\{2\left(\frac{n}{\tilde{\lambda}} \pm \tilde{\nu}\,\tau\right) - \frac{E}{K}\right\}\right] \tag{7.10}$$

where

$$2K\tilde{\nu} = \sqrt{\frac{ab}{m}}\left(\frac{1}{\mathrm{sn}^2\,(2K/\tilde{\lambda})} - 1 + \frac{E}{K}\right)^{-1/2}.$$

In these expressions, K and E are complete elliptic integrals of the 1st and 2nd kinds, respectively, defined with modulus κ in the range $0 < \kappa \leqslant 1$.

Assuming a small κ, we can use approximately $E/K \simeq 1 - \kappa^2/2$, so that the above dn^2–function behaves like a sinusoidal wave. Thus, for a small displacement r_n, (7.10) can be expanded to

$$r_n \simeq -\frac{\tilde{\omega}^2\kappa^2}{8ab}\cos\left(\tilde{\omega}\,\tau \pm \frac{2\pi n}{\tilde{\lambda}}\right) \quad \text{where} \quad \tilde{\omega} = 2\pi\tilde{\nu} = \frac{\sqrt{ab/m}}{\tilde{\lambda}}. \tag{7.11}$$

On the other hand, the quantity in the square bracket of (7.10) can be expressed by Fourier's expansion as

$$\begin{aligned} \mathrm{dn}^2\,(2\pi K) - \frac{E}{K'} &= \frac{\pi^2}{K^2}\sum_{\ell=-\infty}^{\infty}\frac{\ell\cos(2\pi\,\ell\,x)}{\sinh(\pi\,\ell\,K'/K)} \\ &= \left(\frac{\pi}{2K'}\right)^2\left[\sum_{\ell=-\infty}^{\infty}\mathrm{sech}^2\left\{\frac{2K}{K'}(x-\ell)\right\}\right] - \frac{\pi}{2KK'} \end{aligned} \tag{7.12}$$

where $K' = K(\kappa')$ and $\kappa' = \sqrt{1-\kappa^2}$, indicating a series of Eckart potentials.

For thermodynamics, (7.12) should be considered for finite crystals at all temperatures, where each term specified by ℓ corresponds to T_ℓ by time–temperature conversion related with soliton number $n_\ell = \tilde{\lambda}\ell$. Hence, these Eckart's potentials form a lattice structure called *Toda's lattice or soliton lattice*, representing the nature of dual B.

7.2 Developing nonlinearity

The Toda potential (7.8*b*) is employed in a convenient form to analyze binary orderings. Flaschka [2] and Lax [3] showed that displacive order variables in a modulated lattice can be composed of *matrix* formulation in Toda's exponential lattice, which is discussed in this section.

7.2.1 Matrix operators for Toda's correlation potentials

In an exponential lattice determined by Toda's potential $\phi(r_n)$ in the form of (7.8*b*), we consider the coordinate $Q_n = r_n$ and its conjugated momentum $P_n = \dot{Q}_n$, so that we have the relation $P_n = e^{-(Q_n - Q_{n-1})} - e^{-(Q_{n+1} - Q_n)}$. Therefore, we define a new set of variables a_n and b_n as the generalized coordinates for the potential $\phi(a_n, b_n)$, i.e.

$$a_n = \frac{1}{2} e^{-\frac{1}{2}(Q_{n-1} - Q_n)} \quad \text{and} \quad b_n = \frac{1}{2} P_n. \tag{7.13a}$$

It is noted that these new variables are defined to correlate Eckart's potential with $\phi(r)$, as indicated by

$$\dot{a}_n = a_n(b_n - b_{n+1}) \quad \text{and} \quad \dot{b}_n = 2(a_{n-1}^2 - a_n^2). \tag{7.13b}$$

Also note that the definition of (P_n, Q_n) is unchanged by reversing the order of recurrence, so we have the relations

$$a_{n+N} = a_n \quad \text{and} \quad b_{n+N} = b_n. \tag{7.13c}$$

Following Lax [3], the Hamiltonian for a_n and b_n is expressed as follows.

$$\boldsymbol{L} = \begin{pmatrix} b_1 & a_1 & & & & & & a_N \\ a_1 & b_2 & & & & & & \\ & & b_{n-1} & a_n & & & & \\ & & a_n & b_n & a_{n+1} & & & \\ & & & a_{n+1} & b_{n+1} & & & \\ & & & & & & b_{N-1} & a_N \\ a_N & & & & & & a_N & b_N \end{pmatrix}. \tag{7.13d}$$

Corresponding to (5.1) in chapter 5, an antisymmetric matrix

$$\boldsymbol{B}_\pm = \begin{pmatrix} 0 & -a_1 & & & & & & \pm a_N \\ a_1 & 0 & & & & & & \\ & & 0 & -a_{n-1} & & & & \\ & & a_{n-1} & 0 & -a_n & & & \\ & & & a_n & 0 & & & \\ & & & & & & 0 & -a_{N-1} \\ \mp a_N & & & & & & a_{N-1} & 0 \end{pmatrix} \tag{7.13e}$$

are written as the operator of binary inversion $a_n \rightleftarrows -a_n$ in the exponential lattice, to develop positive and negative eigenvalues of $\boldsymbol{L}$.

The equation of motion for nonlinear development under a constant eigenvalue of $\boldsymbol{L}$ can then be written as

$$\frac{\partial \boldsymbol{L}}{\partial \tau} = \boldsymbol{B}_{\pm}\boldsymbol{L} - \boldsymbol{L}\boldsymbol{B}_{\pm} \tag{7.13f}$$

where τ is a temporal variable for development, representing lattice modulation. We call (7.13*f*) Lax' development equation in Toda's lattice, where the transition uncertainty due to inversion can be mathematically avoided, as explained by the unitary matrix $\boldsymbol{U}$ that is defined by

$$\frac{\partial \boldsymbol{U}(t)}{\partial t} = \boldsymbol{B}_{\pm}\boldsymbol{U}(t), \qquad \boldsymbol{U}(0) = \mathbf{1} \tag{7.13g}$$

and t is the *real time*. Then, by writing $\frac{\partial U^{-1}}{\partial t} = -\boldsymbol{U}^{-1}\boldsymbol{B}_{\pm}$, we obtain the expression

$$\boldsymbol{U}\,\boldsymbol{U}^{-1} = \boldsymbol{U}^{-1}\boldsymbol{U} = \mathbf{1} \quad \text{and} \quad \frac{\partial}{\partial t}(\boldsymbol{U}^{-1}\boldsymbol{L}_{\pm}\boldsymbol{U}) = 0,$$

provided that these two timescales are not the same. Thus, we had better write

$$\frac{\partial}{\partial t}\left\{\boldsymbol{U}^{-1}(t)\boldsymbol{L}(\tau)\boldsymbol{U}(t)\right\} = 0, \tag{7.13h}$$

signifying that Lax' Hamiltonian $\boldsymbol{L}(\tau)$ represents a *conservative system*.

At any rate, we have to solve wave equations $\boldsymbol{L}(\phi)\psi(\phi) = \lambda(\phi)\psi(\phi)$ for eigenvalues $\lambda(\phi)$, which is the same as $\lambda(0)$.

The matrix formulation is a logical method for collective motion of order variables in finite crystals. However, to obtain eigenvalues of Lax' Hamiltonian (7.13*a*), we need to solve the secular equation

$$\det|\lambda \boldsymbol{e} - \boldsymbol{L}| = 0, \tag{7.14}$$

where $\boldsymbol{e}$ is the unit-matrix. The eigenvalues can be obtained as real roots of algebraic equation

$$\lambda^N + c_1\lambda^{N-1} + c_2\lambda^{N-2} + \cdots + c_{N-1}\lambda + c_N = 0$$

where $c_1, c_2, \ldots, c_N$ are functions of a_n and b_n, the roots are determined as

$$\lambda = \lambda_1, \lambda_2, \ldots, \lambda_N$$

and all are constant of temperature, but associated with the potential $\phi_n(\tau)$ that is determined by $\frac{\partial \phi_n(\tau)}{\partial \tau} = 0$, i.e. $\phi_n(r_0)$ is regarded as a function of temperature after time–temperature conversion.

7.2.2 Finite periodic lattice

We are solving (7.13*f*) for finite crystals, however the boundary condition should by specified precisely for Toda's lattice. While antisymmetric, the matrix operators $\boldsymbol{L}$ and $\boldsymbol{B}_\pm$ must be subjected to boundaries characterized by unavoidable ambiguities. Nevertheless, if these operators are commutable, such ambiguities as $\Delta\boldsymbol{B}_\pm$ are not explicit in (7.13*f*).

Therefore, we consider for boundaries with respect to the number N to represent a *finite* one-dimensional crystal that is ended by $-\frac{N}{2}$ and $+\frac{N}{2}$ on the left and right *surfaces*, respectively; which is equivalent with a single Brillouin zone of a uniform *infinite* crystal.

For such a finite crystal, Flaschka [2] wrote

$$L\psi(n) = a_{n-1}\psi(n-1) + b_n\psi(n) + a_n\psi(n+1) = \lambda\psi(n) \tag{7.15a}$$

and

$$\frac{d\psi(n)}{d\tau} = \boldsymbol{B}_+\psi(n) = a_{n-1}\psi(n-1) - a_n\psi(n+1), \tag{7.15b}$$

where $n = -\frac{N}{2}, \ldots, -2, -1, 0, 1, 2, \ldots, \frac{N}{2}$ and $\lambda = \text{const.}$

In the region for $N \to \infty$, where interactions are limited to nearest neighbors, we can assume $Q_{n+1} - Q_n = 0$ and $P_n = 0$ and hence $a_n = \frac{1}{2}$ and $b_n = 0$ at a long distance $|n| \gg 1$, allowing to express the wave function $\psi(n)$ in the *asymptotic form*

$$\psi(n) \sim e^{\pm i(\tilde{\omega}\tau \pm qn)} = e^{\pm i\tilde{\omega}\tau}(e^{\pm iq})^n = e^{\pm i\tilde{\omega}\tau}z^n \quad \text{where} \quad z = e^{iq}, \tag{7.16a}$$

which should satisfy the asymptotic equations

$$\frac{1}{2}\{\psi(n-1) + \psi(n+1)\} = \lambda\psi(n) \quad \text{and} \quad \frac{d\psi(n)}{d\tau} = \frac{1}{2}\{\psi(n-1) - \psi(n+1)\},$$

and obtain

$$\lambda = \frac{z + z^{-1}}{2} = \cos q \quad \text{and} \; -\tilde{\omega} = \frac{z - z^{-1}}{2i} = \sin q. \tag{7.16b}$$

Here, z is a parameter for the singularity related to q, playing a dominant role in the following discussions.

The wave function is expressed by (7.16*a*), for which the *nonlinear phase* is essentially determined by z. Since n is an integer, q should be *real wavevector* in the range $0 < q \leqslant 1$, expressing a mesoscopic wave. Further, the eigenvalue λ has a critical value $\lambda_c = 1$ thermodynamically, and associated with the frequency $\tilde{\omega}$ for $0 < \lambda \leqslant 1$. Otherwise for $\lambda > 1$, we have $|z| \neq 1$, and no possibility for propagation.

On the other hand, in the active region where $a_n \neq 1/2$ and $b_n \neq 0$, waves are scattered and captured by potentials depending on the value of z, called the *scattering region*. In fact, the potential is determined from the given asymptotic waves, and the mathematical process is called *inverse scatterings*. For the inverse scattering, (7.15*a*) is expressed as

$$a_{n-1}\psi(n-1) + b_n\psi(n) + a_n\psi(n+1) = \frac{1}{2}(z + z^{-1})\psi(n). \tag{7.17a}$$

It is further noted that mathematically the eigenvalues can also be permitted for the negative range $-1 \leqslant \lambda \leqslant 0$. It is physically significant to support the processes scattering and capture by Toda's potentials, as will be discussed in chapter 8.

We therefore consider an asymptotic wave function

$$\psi(n) = \varphi(n, z)e^{i\tilde{\omega}\tau} \quad \text{where} \quad \varphi(n, z) = \sum_{n'=n-1}^{n'=n+1} K(n, n')z^{n'}, \tag{7.15c}$$

instead of (7.15*a*), to replace $\psi(n)$ in (7.15*b*). Thus, comparing coefficients of z^{n-1}, z^n and z^{n+1} terms, we derive the following relations

$$a_{n-1}K(n-1, n-1) = \frac{1}{2}K(n, n),$$

$$a_{n-1}K(n-1, n) + b_nK(n, n) = \frac{1}{2}K(n, n+1),$$

$$a_{n-1}K(n-1, n+1) + a_nK(n+1, n+1) + b_nK(n, n+1) = \frac{1}{2}\{K(n, n) + K(n, n+2)\},$$

$$a_{n-1}K(n-1, n+2) + a_nK(n+1, n+2) + b_nK(n, n+2) = \frac{1}{2}\{K(n, n+1) + K(n, n+3)\},$$

…………………

Solving the first two relations for a_{n-1} and b_n, we obtain

$$a_{n-1} = \frac{K(n, n)}{2K(n-1, n-1)} \quad \text{and} \quad b_n = \frac{K(n, n+1)}{2K(n, n)} - a_{n-1}\frac{K(n-1, n)}{K(n, n)},$$

therefore

$$b_n = \frac{K(n, n+1)}{2K(n, n)} - \frac{K(n-1, n)}{2K(n-1, n-1)}.$$

Returning to the definitions P_n and Q_n in (7.13*a*), it is now clear that we have the relations:

$$\left\{\frac{K(n, n)}{K(n-1, n-1)}\right\}^2 = e^{Q_{n-1}-Q_n} \quad \text{and} \quad P_n = s_{n-1} - s_n,$$

hence writing

$$s_n = -\frac{K(n, n+1)}{K(n, n)} \quad \text{and} \quad \dot{s}_n = e^{Q_n - Q_{n+1}} - 1.$$

Thus, we confirm that the matrix formulation with the exponential correlations is compatible with the soliton theory characterized by discrete number n.

7.3 Conversion to Korteweg–deVries' lattice potential

From Toda's lattice discussed in this chapter, the exponential correlation potential emerges as algebraically analyzable, but exhibiting the same feature as derived from Korteweg–deVries' equation. Dealing with finite lattice displacements, solitons should be an obvious consequence of Toda's theory. In this section, we show that the Toda lattice can be transformed to the KdV equation in an approximate manner, keeping the advantage of integrability in Toda's theory [1].

Writing the correlation potential as $\phi_n(\tau)$, Toda's equation of motion can be expressed as

$$\frac{d^2}{d\tau^2}\ln(1+\phi_n) = \phi_{n+1} + \phi_{n-1} - 2\phi_n, \tag{7.16}$$

and we carry out the space–time conversion $n \to x$ and $\tau \to h^2 t$, where h is a positive parameter in the range $0 \leqslant h^2 \leqslant 1$, writing $\phi_n(x, t) = h^2 u_n(t)$. In other words, the conversion is performed as

$$x = hu - \left(\frac{1}{h^2} - h^2\right)t \quad \text{and}$$

$$\begin{aligned}&\left\{\frac{\partial}{\partial\tau} - \left(\frac{1}{h^2} - h^2\right)\frac{\partial}{\partial x}\right\}^2 \ln\left\{1 + h^2 u(x, \tau)\right\}\\ &\quad= \frac{1}{h^2}\left\{u(x+h, \tau) + u(x-h, \tau) - 2u(x, \tau)\right\}.\end{aligned} \tag{7.17}$$

In this expression, if $h = 1$, (7.17) is the same as (7.16), while (7.15) represents Toda's lattice with $h \neq 1$. However, a limiting case of $h \to +0$, we have

$$\frac{\partial}{\partial x}\left\{-2\frac{\partial u}{\partial x} - \frac{1}{2}\left(\frac{\partial u}{\partial x}\right)^2\right\} = \frac{1}{12}\frac{\partial^4 u}{\partial x^4}.$$

In (7.17), the integrable part can be expressed by a KdV equation,

$$\frac{\partial u}{\partial \tau} + \frac{1}{2}u\frac{\partial u}{\partial x} + \frac{1}{24}\frac{\partial^3 u}{\partial x^3} = 0. \tag{7.18a}$$

Thus, (7.17) and (7.18) of a Toda's lattice is consistent with the KdV equation (7.16) in the limit $h \to +0$, permitting to interpret this approach as a process for two Eckart peaks to collide at a crossing point in a phase diagram.

In general, Toda's potential $\phi_n(\tau)$ derived from (7.13) can be converted *in phase* to KdV's equation in the limit of $h \to +0$, appearing in general form

$$\frac{\partial u_j}{\partial \tau} - 6u_j\frac{\partial u_j}{\partial x} + \frac{\partial^3 u_j}{\partial x^3} = 0 \quad \text{and} \quad u_n = \sum_{j=1}^{j=n} u_j, \tag{7.18b}$$

where the potential $u_n(\tau)$ is composed by the collision of two coherent potentials, and ϕ is the total potentials at the crossing point (x, τ).

When two soliton peaks are crossing over, each of Lax' Hamiltonian equations can be written for the wave functions ϕ_j as

$$\boldsymbol{L}\,\varphi_j = -q_j^2\,\varphi_j \quad \text{and} \quad \boldsymbol{L} = -\frac{\partial^2}{\partial x^2} - u$$

which is modified by the development equation

$$\frac{\partial \varphi_j}{\partial \tau} = \boldsymbol{B}\,\varphi_j \quad \text{where} \quad \boldsymbol{B} = -4\frac{\partial^3}{\partial x^3} - 6u\frac{\partial}{\partial x} - 3\frac{\partial u}{\partial x}.$$

Normalized as usual to 1, here we have the relation $u_j = 4q_j\phi_j^2$.

For two solitons $j = 1$ and 2 to collide as illustrated in figures 6.1(*a*) and (*b*) of chapter 6, we consider the normalized wave function expressed as $\psi_j = c_1\phi_1 + c_2\varphi_2$ where $c_1^2 + c_2^2 = 1$, solving the impact problem for the collision energy at minimum level, i.e. $\boldsymbol{BL} = \boldsymbol{LB}$. Therefore, eigenfunctions φ_1 and φ_2 can be expressed as follows.

$$\varphi_1 = \frac{1}{\Delta}\begin{vmatrix} c_1 e^{\psi_1} & \dfrac{c_1c_2e^{\psi_1+\psi_2}}{q_1+q_2} \\ c_2e^{\psi_2} & 1+\dfrac{c_1c_2e^{2\psi_2}}{q_1+q_2} \end{vmatrix} \quad \text{and} \quad \varphi_2 = \frac{1}{\Delta}\begin{vmatrix} 1+\dfrac{c_1c_2e^{2\psi_1}}{q_1+q_2} & c_1e^{\psi_1} \\ \dfrac{c_1c_2e^{\psi_1+\psi_2}}{q_1+q_2} & c_2e^{\psi_2} \end{vmatrix}$$

where

$$\Delta = \det\left|\delta_{1,2} + \frac{c_1c_2e^{\psi_1+\psi_2}}{q_1+q_2}\right|, \quad \psi_1 = q_1x_1 - 4\kappa_1\tau_1 \quad \text{and} \quad \psi_2 = q_2x_2 - 4\kappa_2\tau_2.$$

These are the same as Bargmann's two-soliton solution discussed in chapter 6, showing the coherent phase $\psi_1 = \psi_2$ at the crossing point.

Exercises

1. Notice that the Toda theorem is consistent with the soliton theory, comprising the Weiss field. If considering a non-thermodynamic perturbation, we may have to discuss effects other than the Toda lattice. Discuss the issue.
2. Despite question 1, section 7.3 indicated that Toda's lattice is compatible with the KdV equation. What is wrong? Discuss this issue.
3. It is notable that Toda introduced the exponential correlation potential on the collective order variable defined by the solution theory. Hence, there should be no conflicts with the soliton theory, and we find it all consistent. Discuss the issue, considering how he discovered such integrable potentials, because it is pedagogically important.

References

[1] Toda M 1987 *Lectures on Nonlinear Lattice Dynamics* (in Japanese) (Tokyo: Iwanami)
Toda M 1974 *Phys. Rep.* **18c1**
Toda M 1989 *Theory of Nonlinear Lattices* 2nd ed (Berlin: Springer)
[2] Flaschka H 1974 *Phys. Rev.* B **9** 1924
[3] Lax P D 1668 *Commun. Pure Appl. Math.* **21** 467

IOP Publishing

Introduction to the Mathematical Physics of Nonlinear Waves
(Second Edition)

Minoru Fujimoto

Chapter 8

Scattering theory of the soliton lattice

Signified by longitudinal and transverse components, nonlinear waves are not so simple as in one-dimension, but exhibit the essential feature described by the phase variable with finite amplitude, together with the phase velocity through medium.

Longitudinal waves provide only an approximate view of propagation, while the strained media is complex but in steady state with respect to the lattice, whereas the transversal components disperse and reflect from surfaces and obstacles in crystals, responsible for possible heat dissipations.

However, in conservative crystals at equilibrium temperatures, the soliton lattice is steadily modulated, nonlinear waves of order variables should be discussed in the field of soliton potentials. In this chapter, we discuss that the scattering theory revealed that the domain wall gives rise to an *inductive reflection and transmission*, for which the Weiss field is not responsible. Instead, the imaginary potential $\pm i\frac{\mathrm{d}u}{\mathrm{d}x}$ are confirmed to play a significant role.

8.1 Elemental waves

8.1.1 Critical fluctuations

As introduced in section 3.1 in chapter 3, we consider *free running waves* in crystals, where the pseudospin vector of order variable $\boldsymbol{\sigma}_q(x, t)$ specified by a specific $\boldsymbol{q}$. Hence, writing a wave equation

$$\frac{\partial^2 \boldsymbol{\sigma}_q}{\partial t^2} - \frac{\partial^2 \boldsymbol{\sigma}_q}{\partial x^2} = 0, \tag{8.1a}$$

the order variable is expressed by $\boldsymbol{\sigma}_q = \boldsymbol{\sigma}_1(x - t) + \boldsymbol{\sigma}_2(x + t)$ where $\boldsymbol{\sigma}_1$ and $\boldsymbol{\sigma}_2$ represent waves propagating toward the $+x$ and $-x$ directions, respectively, at a speed 1.

doi:10.1088/978-0-7503-3759-5ch8

Writing $u = \frac{\partial \sigma_q}{\partial t}$ and $v = \frac{\partial \sigma_q}{\partial x}$, (8.1) can be factorized as

$$\left(\frac{\partial}{\partial t} + \frac{\partial}{\partial x}\right)(v - u) = 0 \quad \text{and} \quad \left(\frac{\partial}{\partial t} - \frac{\partial}{\partial x}\right)(v + u) = 0, \tag{8.1b}$$

which are illustrated as shown in the x-t plane in figure 8.1. Here $v \mp u$ are unchanged in phase shifts $\xi = x + t$ and $\eta = x - t$, respectively, known as *Riemann's invariants.* We can write $v - u \propto \xi$ and $v + u \propto \eta$ for a simple case, as illustrated by groups of parallel lines in figure 8.1(*b*), however the phase relations for nonlinear waves should be modified by the media, therefore we express the relation as

$$v - u = r(\xi) \quad \text{and} \quad v + u = s(\eta) \tag{8.1c}$$

which are functions of the media.

If ξ and η are microscopic quantities, we should consider quantum-mechanical uncertainties $\Delta\xi$ and $\Delta\eta$ at the threshold for nonlinearity, these straight parallel lines in figure 8.1(*b*) must be replaced by *classical bifurcation.* In any case,

$$\begin{aligned} \sigma_1(x - t) &= \sigma_0(\eta) + \int_{-\Delta\xi}^{+\Delta\xi} v_0(\xi)\, \mathrm{d}\xi \\ \text{and} \quad \sigma_2(x + t) &= \sigma_0(\xi) + \int_{-\Delta\eta}^{+\Delta\eta} u_0(\eta)\, \mathrm{d}\eta. \end{aligned} \tag{8.1d}$$

In practice, such an excitation is discontinuous, the waves $\sigma_1(x - t)$ and $\sigma_2(x + t)$ may not be mixed evenly, depending on the initial situation. In figure 8.1(*a*), the discontinuity occurs in the region around $(x_0,\ t_0)$. At time t later than t_0, these elemental waves are separated from each other, which is a typical feature of nonlinear waves.

8.1.2 Matrix form for nonlinear development

The nonlinear equation can be formulated in the matrix form for two components u and v, constituting the *origin of nonlinearity*.

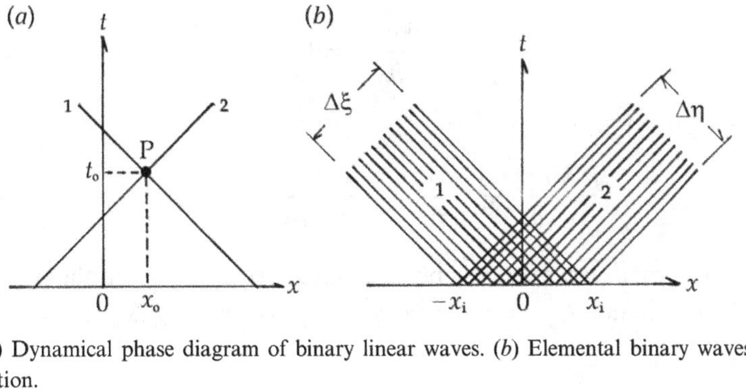

Figure 8.1. (*a*) Dynamical phase diagram of binary linear waves. (*b*) Elemental binary waves of nonlinear wave propagation.

Defining the unit matrix $U_o = \binom{u}{v}$ and $U_o^{-1} = (u\ v)$, where the latter is the transposed row matrix of the former, we can then write

$$\frac{\partial U_o^{-1}}{\partial t} + (A_o)\frac{\partial U_o}{\partial x} = 0 \quad \text{where} \quad (A_o) = \begin{pmatrix} 0 & 1 \\ -1 & 0 \end{pmatrix} \tag{8.2}$$

is a matrix to connect the two components u and v. Using unit vectors defined by $\varepsilon_1^{-1} = (1, 0)$ and $\varepsilon_2^{-1} = (0, 1)$ for the matrix (A_o), we have the relations $\varepsilon_{1,2}^{-1}(A_o) = \pm\, \varepsilon_{1,2}^{-1}$ and $(A_o)\varepsilon_{1,2} = \varepsilon_{1,2}$, indicating that (A) has eigenvalues $\pm$ 1.

Therefore, the matrix $U^{-1} = (\sigma_1\varepsilon_1^{-1},\ \sigma_2\varepsilon_2^{-1})$ for the complex order variable can be expressed as

$$\frac{\partial\sigma_1}{\partial t} + (-1)\frac{\partial\sigma_1}{\partial x} = 0 \quad \text{and} \quad \frac{\partial\sigma_2}{\partial t} + (+1)\frac{\partial\sigma_2}{\partial x} = 0, \tag{8.3}$$

describing component waves σ_1 and σ_2 for eigenvalues -1 and $+1$, respectively.

The above formulation can be extended to systems composed by more-than two components, as already discussed in chapter 7. Nevertheless, in this section, we need to discuss heat dissipation of the wave motion, as an example of matrix method applied to macroscopic surroundings in the following.

In practice, crystals are surrounded by outside air, which is considered as the heat reservoir. However, inside is composed by the soliton lattice that must be another heat reservoir. Accordingly, we must regard the crystal lattice for the effective reservoir for excessive deformation energy created by the soliton lattice.

Nonetheless, in field theory solitons are expressed by a system of freely moving pulses of potential energies, it is logical to consider it like an ideal gas of particles characterized by effective pressure and temperature.

Therefore, thermodynamically we need to consider a change in the internal energy for produced entropy ΔS, which is expressed by $T\Delta S$ under constant volume condition. In this case, an effective internal pressure-change Δp is responsible for $\Delta S = -\frac{R}{p}\Delta p$, under a constant T, which can be detected with externally applied pressure $-\Delta p_{\text{ext}}$.

Propagating nonlinear waves are described by the *continuity equation of density and current*, i.e. density $\rho(\boldsymbol{r},\ t)$ and speed $v(\boldsymbol{r},\ t)$ expressed in this case

$$\frac{\partial\rho}{\partial t} + \frac{\partial(\rho v)}{\partial x} = 0, \tag{8.4a}$$

and the propagation equation

$$\frac{\partial v}{\partial t} + v\frac{\partial v}{\partial x} = -\frac{1}{\rho}\frac{\partial(\Delta p)}{\partial x}. \tag{8.4b}$$

For an isotropic medium, Δp can be calculated from the formula $p = \text{const.} \times \rho^\gamma$ where $\gamma = C_p/C_V$, hence $\Delta p = \text{const.} \times \gamma\rho^{\gamma-1}\Delta\rho$; or writing $\Delta p/\Delta\rho = a^2$ where a is sound velocity, we have the relation $a^2 = \text{const.} \times \gamma\, \rho^{\gamma-1}$ to be used. Therefore, we re-express (8.4*a*) and (8.4*b*) as

$$\frac{\partial \rho}{\partial t} + v\frac{\partial \rho}{\partial x} + \rho\frac{\partial v}{\partial x} = 0 \quad \text{and} \quad \frac{\partial v}{\partial t} + v\frac{\partial v}{\partial x} + \frac{a^2}{\rho}\frac{\partial \rho}{\partial x} = 0. \tag{8.4c}$$

These can be transformed in matrix form for $U = \begin{pmatrix} \rho \\ v \end{pmatrix}$ and $U^{-1} = (\rho,\ v)$ to be described as

$$\frac{\partial U}{\partial t} + (A)\frac{\partial U}{\partial x} = 0 \quad \text{where} \quad (A) = \begin{pmatrix} v & \rho \\ a^2\rho^{-1} & v \end{pmatrix}. \tag{8.4d}$$

Diagonalizing the matrix (A), we have

$$(A)\begin{pmatrix} \rho_0 \\ v_0 \end{pmatrix} = \begin{pmatrix} v + a & 0 \\ 0 & v - a \end{pmatrix}\begin{pmatrix} \rho_0 \\ v_0 \end{pmatrix}.$$

Noting that $\Delta S \propto \Delta p = 0$, if $a = 0$, these eigenvalues v represents waves for energy flow to surroundings to be determined by the temperature T.

8.2 Scattering theory: dissipation, reflection, and transmission

8.2.1 Elemental waves

In the presence of a localized potential $V(x)$ at a Galilean coordinate x, regarded as a dynamical phase of $(x,\ t)$, the wave equation

$$\frac{d^2\sigma_q}{dt^2} + (q^2 - V(x))\sigma_q = 0 \tag{8.5a}$$

is considered for the complex wave function $\psi_q = \psi_q' + i\,\psi_q''$ for the density $\sigma_q = \psi_q'\psi_q''$. Therefore, for the real potential, expressing (8.5*a*) for real and imaginary parts of ψ_q, we have

$$\frac{d^2\psi_q'}{dx^2} + q^2\psi_q' = V(x)\psi_q' \quad \text{and} \quad \frac{d^2\psi_q''}{dx^2} = V(x)\psi_q''.$$

Multiplying by $\frac{d\psi_q''}{dx}$ and $\frac{d\psi_q'}{dx}$ on each of these equations, respectively, and subtract then integrate it over $-\infty < x < +\infty$, we obtain

$$\left(\psi_q''\frac{d\psi_q'}{dx} - \psi_q'\frac{d\psi_q''}{dx}\right)_{-\infty}^{+\infty} = 2q(-q)\int_{-\infty}^{+\infty}\psi_q'\psi_q''\, dx = 0, \tag{8.5b}$$

indicating the presence of a singularity at $x = 0$.

Further, the wave function is determined by a linear combination of $e^{\pm iqx}$, approaching $x \to \pm\infty$, perturbed with amplitudes $V(x_\pm)f_{\pm q}(x_\pm)$, so that

$$\psi_q = V(x_+)f_q(x_+)e^{iqx} + V(x_-)f_{-q}(x_-)e^{-iqx} \tag{8.6}$$

where

$$f_q(x_+) = e^{iqx} - \frac{1}{q}\int_{x_+}^{+\infty} V(x) f_q(x_+) \sin q(x - x_+) \mathrm{d}x$$

and

$$f_{-q}(x_-) = e^{-iqx} + \frac{1}{q}\int_{-\infty}^{x_-} V(x) f_{-q}(x_-) \sin q(x - x_-) \mathrm{d}x,$$

expressing elemental waves to interact with the potential $V(x)$.

Using approximated (8.6), we can express elemental waves (ε_1, $\varepsilon_\perp$) propagating from left to right as

$$\varepsilon_> = c_{11}(q)e^{iqx} + c_{1\perp}(-q)e^{-iqx} \quad \text{and} \quad \varepsilon_< = e^{-iqx}; \tag{8.7a}$$

and from right to left as

$$\varepsilon_> = e_{iqx} \quad \text{and} \quad \varepsilon_< = c_{\perp 1}(q)e^{iqx} + c_{\perp\perp}(-q)e^{-iqx}, \tag{8.7b}$$

where these coefficients c_{11}, $c_{\perp\perp}$, $c_{1\perp}$ and $c_{\perp 1}$ are related by

$$c_{11}(q) f_q(x_+) + c_{1\perp}(-q) f_{-q}(x_-) = f_{-q}(x_-) \tag{8.7c}$$

and

$$c_{\perp 1}(-q) f_{-q}(x_-) + c_{\perp\perp}(q) f_q(x_+) = f_q(x_+). \tag{8.7d}$$

These (8.7*c*) and (8.7*d*) determine incident waves from the right (R) and the potential at $x = 0$ (L), as shown in figure 8.2(*a*). It is noted that among these coefficients we have relations

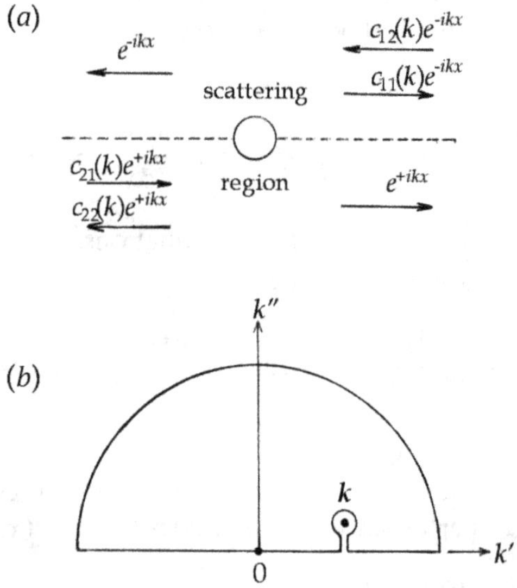

Figure 8.2. (*a*) Reflection and transmission of elemental waves. (*b*) Cauchy's diagram in *k*-space.

$$c_{11}(q)c_{\perp\perp}(q) + c_{1\perp}(q)c_{\perp 1}(-q) = 1 \quad \text{and} \quad c_{11}(q)c_{\perp 1}(q) + c_{\perp\perp}(-q)c_{1\perp}(q) = 0.$$

Accordingly, we can define reflection and transmission factors by

$$R_{\mathrm{R}}(q) = \frac{c_{11}(q)}{c_{1\perp}(q)}, \quad R_{\mathrm{L}}(-q) = \frac{c_{\perp\perp}(-q)}{c_{1\perp}(-q)}; \quad T_{\mathrm{R}}(q) - \frac{1}{c_{1\perp}(q)}, \qquad (8.7e)$$
$$T_{\mathrm{L}}(-q) = \frac{1}{c_{\perp 1}(-q)}.$$

8.2.2 Reflection and transmission of two-component waves

For (8.6) however, the wave function should represent two-component wave $(\sigma_1, \ \sigma_\perp)$, in a Weiss field proportional to $\sigma_1^2 + \sigma_\perp^2$, (8.6) should be replaced by

$$\frac{\mathrm{d}\sigma_1}{\mathrm{d}x} - iu(x)\sigma_1 = -q\ \sigma_\perp \quad \text{and} \quad \frac{\mathrm{d}\sigma_\perp}{\mathrm{d}x} + iu(x)\sigma_\perp = q\ \sigma_1 \qquad (8.8a)$$

where $u(x) = -\frac{K(x)}{v}$, describing the spatial motion.

Hence, we define matrices

$$\boldsymbol{U}_q^{-1} = (\sigma_1, \ \sigma_\perp), \quad U_{\mathrm{Weiss}} = \begin{pmatrix} +u\ \sigma_\perp \\ -u\ \sigma_1 \end{pmatrix} \quad \text{and} \quad (A) = \begin{pmatrix} iq & 0 \\ 0 & -iq \end{pmatrix},$$

and (8.8) can be expressed as

$$\frac{\mathrm{d}\boldsymbol{U}_q^{-1}}{\mathrm{d}x} + (A)\boldsymbol{U}_q^{-1} = U_{\mathrm{Weiss}}. \qquad (8.8b)$$

Noticing that the homogeneous equation of (8.8*b*) has independent solutions $(e^{-iqx}, \ e^{iqx})$, and their perturbed solutions are written as

$$\left(\Psi_{-q}(x_-)e^{-iqx}, \ \Psi_q(x_+)e^{iqx}\right) \quad \text{and} \quad \left(\Phi_q(x_+)e^{-iqx}, \ \Phi_{-q}(x_-)e^{-iqx}\right) \qquad (8.8c)$$

where $\Psi_{-q}(x_-)$, $\Psi_q(x_+)$ and $\Phi_q(x_+)$, $\Phi_{-q}(x_-)$ are to be determined with respect to $\sigma_1/\sigma_\perp$, respectively, however depending on damping in thermodynamic environment, and unequal to 1 in general. Consequently, these factors are expressed for the first and second cases of figure 8.2(*a*) as listed in the following.

$$\Psi_{-q}(x) = e^{-iqx} - \int_{x_+}^{\infty} \Psi_{-q}(x_+)\,u(x_+)\,e^{+iq(x-x_+)}\mathrm{d}x_+ \quad \text{and} \quad \Psi_{-q}(x)$$
$$= - \int_{-\infty}^{x_-} \Psi_{-q}(x_-)\,u(x_-)e^{-iq(x-x_-)}\mathrm{d}x_-\boldsymbol{Z}$$
$$\Phi_q(x) = - \int_{x_+}^{\infty} \Phi_q(x_+)\,u(x_+)e^{-iq(x_+-x)}\mathrm{d}x_+ \quad \text{and} \quad \Phi_q(x)$$
$$= e^{iqx} + \int_{-\infty}^{x_-} \Phi_q(x_-)u(x_-)\,e^{iq(x-x_-)}\mathrm{d}x_-.$$

These factors are essential for conversion $\boldsymbol{q} \rightleftarrows -\boldsymbol{q}$ of n-coherent elemental solitons, hence related as

$$\Phi_q(x) = c_{11}(\boldsymbol{q})\Psi_q(x) + c_{1\perp}(\boldsymbol{q})\bar{\Psi}_q(x), \quad \text{and} \quad \Psi_q(x) = c_{\perp 1}(\boldsymbol{q})\bar{\Phi}_q(x) + c_{\perp\perp}(\boldsymbol{q})\Phi_q(x)$$
$$\bar{\Phi}_q(x) = c_{11}(-\boldsymbol{q})\bar{\Psi}_q(x) - c_{1\perp}(-\boldsymbol{q})\Psi_q(x) \quad \text{and} \quad \bar{\Psi}_q(x) = -\, c_{\perp 1}(-\boldsymbol{q})\Phi_q(x) + c_{\perp\perp}(-\boldsymbol{q})\bar{\Phi}_q(x)$$

where

$$c_{11}(\boldsymbol{q}) = -\, c_{\perp\perp}(-\boldsymbol{q}), \quad c_{1\perp}(-\boldsymbol{q}) = c_{1\perp}(\boldsymbol{q}), \quad c_{1\perp}(\boldsymbol{q}) = -\, c_{\perp 1}(\boldsymbol{q}), \quad \text{and} \quad c_{\perp\perp}(\boldsymbol{q}) = c_{11}(-\boldsymbol{q}).$$

We therefore obtain the relations

$$c_{11}{}^*(\boldsymbol{q}) = c_{\perp\perp}(-\boldsymbol{q}), \quad c_{1\perp}{}^*(\boldsymbol{q}) = c_{\perp 1}(-\boldsymbol{q}),$$

and

$$c_{11}(\boldsymbol{q})c_{\perp\perp}(\boldsymbol{q}) - c_{1\perp}(\boldsymbol{q})c_{\perp 1}(-\boldsymbol{q}) = 1,$$
$$c_{11}(\boldsymbol{q})c_{\perp 1}(\boldsymbol{q}) + c_{1\perp}(\boldsymbol{q})c_{\perp\perp}(-\boldsymbol{q}) = 0$$
$$c_{11}(\boldsymbol{q})c_{\perp 1}(-\boldsymbol{q}) + c_{1\perp}(-\boldsymbol{q})c_{\perp\perp}(-\boldsymbol{q}) = 0.$$

The above linear combination of $(\sigma_1, \ \sigma_\perp)$ satisfies the same for elemental $(\varepsilon_1, \ \varepsilon_2)$, (8.7*e*) can be expressed as reflection and transmission factors, showing the relation

$$R_R(\boldsymbol{q})T(-\boldsymbol{q}) - R_L(-\boldsymbol{q})T(\boldsymbol{q}) = 0, \tag{8.9a}$$

confirming the relation

$$|T(\boldsymbol{q})|^2 = 1 + |R_R(\boldsymbol{q})|^2 = 1 + |R_L(\boldsymbol{q})|^2 \tag{8.9b}$$

for the two-components variable $(\sigma_1, \ \sigma_\perp)$.

However, to calculate the values of these reflection and transmission factors, we need to obtain poles of the Cauchy integral as illustrated in complex plane in figure 8.2(*b*).

8.2.3 Singularity of reflection and transmission

Due to (8.5*b*) the singularity occurs as related to the component $\sigma_\perp$ responding to the potential $\pm\frac{\mathrm{d}u}{\mathrm{d}x}$, hence expressed by specially written as $\frac{\mathrm{d}c_{1\perp}}{\mathrm{d}x} = \check{c}_{1\perp}(\pm\boldsymbol{q}_\perp)$; i.e.

$$\check{c}_{1\perp}(\pm\boldsymbol{q}_\perp) = c_{\perp\perp}\left(W\left[\frac{\mathrm{d}\Psi}{\mathrm{d}x}, \bar{\Psi}\right]\right) + c_{11}\left(W\left[\frac{\mathrm{d}\bar{\Phi}}{\mathrm{d}x}, \Phi\right]\right) = -2i\int_{-\infty}^{+\infty} \Phi_{\pm q}(x)\Psi_{\mp q}(x)\mathrm{d}x.$$

Accordingly, we obtain

$$R_R(\boldsymbol{q}_\perp) = -i\frac{c_{\perp\perp}(\boldsymbol{q}_\perp)}{\check{c}_{1\perp}(\boldsymbol{q}_\perp)} = \frac{1}{2}\int_{-\infty}^{+\infty} \Phi_{q_\perp}(x)\bar{\Phi}_{q_\perp}\mathrm{d}x$$

and

$$R_L(\boldsymbol{q}_\perp) = -i\frac{c_{11}(\boldsymbol{q}_\perp)}{\check{c}_{1\perp}(\boldsymbol{q}_\perp)} = \frac{1}{2}\int_{-\infty}^{+\infty} \Psi_{q_\perp}(x)\bar{\Psi}_{q_\perp}(x)\mathrm{d}x.$$

For the above expressions, we can use Bargmann's one-soliton solution discussed in chapter 6, that is $\sigma_1 = e^{iqx}F(x)$ where $F(x) = 2q_1 + ia(x)$, to derive the potential

$$V_1(x) = -2\mu^2\mathrm{sech}^2(\mu x - \phi)$$

where the constant μ is determined by the relation $\frac{\mathrm{d}a}{\mathrm{d}x} + \frac{1}{2}a^2 = 2\mu^2$, and ϕ by $a(x) = 2\frac{\mathrm{d}\ln w}{\mathrm{d}x}$, to obtain $\frac{\mathrm{d}^2 w}{\mathrm{d}x^2} - \mu^2 w = 0$ for $w = \alpha\, e^{-\mu x} + \beta e^{+\mu x}$ and $\phi = \frac{1}{2}\ln\frac{\beta}{\alpha}$.

Accordingly, we have $V_1(x) = -u(x)^2$ and $u(x) = \sqrt{2}\,\mu\mathrm{sech}(\mu x - \phi)$, which can be used for $\sigma_\perp(y)$ to propagate along the direction y perpendicular to the x-axis.

Hence, along the y-axis, we calculate

$$\Psi_{-q_\perp}(y) = e^{-iq_\perp y}\frac{2iq_\perp + \mu\tanh(\mu y - \phi)}{-2iq_\perp - \mu}, \quad \Psi_{q_\perp}(y) = \pm\,\mu\, e^{iq_\perp y}\frac{\mathrm{sech}(\mu y - \phi)}{2iq_\perp - \mu},$$

$$\Phi_{-q_\perp}(y) = \pm\,\mu e^{iq_\perp y}\frac{\mathrm{sech}(\mu y - \phi)}{-2iq_\perp - \mu} \quad \text{and} \quad \Phi_{q_\perp}(y) = \mp\, e^{-iq_\perp y}\frac{2iq_\perp - \mu\tanh(\mu y - \phi)}{2iq_\perp - \mu},$$

for which

$$\check{c}_{1\perp} = \frac{2q_\perp - i\mu}{2q_\perp + i\mu}. \tag{8.10a}$$

Accordingly, the pole of singularity can be specified by

$$\Psi_{-q_\perp}\left(y, \frac{i\mu}{2}\right) = \frac{1}{2}e^{-\mu y/2}\mathrm{sech}(\mu y), \quad \Psi_{q_\perp}\left(y, \frac{i\mu}{2}\right) = \mp\frac{1}{2}e^{\mu y/2}\mathrm{sech}(\mu y),$$

$$\Phi_{-q_\perp}\left(y, \frac{i\mu}{2}\right) = \mp\frac{1}{2}e^{-\mu y/2}\mathrm{sech}(\mu y) \quad \text{and} \quad \Phi_{q_\perp}\left(y, \frac{i\mu}{2}\right) = \frac{1}{2}e^{\mu y/2}\mathrm{sech}(\mu y).$$

and the related constant is

$$\check{c}_{1\perp}\left(\frac{i\mu}{2}\right)_R = \check{c}_{1\perp}\left(\frac{i\mu}{2}\right)_L = \mp\,\mu. \tag{8.10b}$$

Figure 8.3 shows the pulsed spectra of sechϕ and sech$^2\,\phi$ near the singular phase.

8.3 Method of inverse scattering

For practical applications of the scattering theory, the significant attempts are to obtain information on the expansion coefficients $K(n, n')$ of (7.15c) of Toda's lattice discussed in chapter 7. This can be achieved mathematically by calculating the lattice potential from scattered waves in the asymptotic form, which is known as inverse scatterings. Notably, there exists a theorem that was worked out by Gel'fand, Levitan and Marchenko (GLM) [1].

According to Toda's theory of soliton lattice, (7.17a) in chapter 7 is a linear equation, so that it is convenient to define reflection and transmission factors for

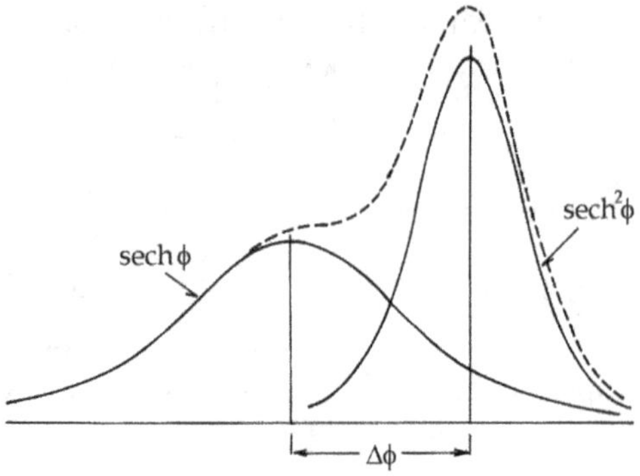

Figure 8.3. Nonlinear potential energy sech$^2\phi$ and nonlinear potential sech ϕ '.

nonlinear waves $\varphi(n, z)$, like in linear waves. Considering z for a complex variable of propagation to $\pm\infty$, we consider asymptotic functions

$$\varphi(n, z) \to z^n, \quad \varphi(n, z^{-1}) \to z^{-n} \quad \text{for} \quad n \to +\infty$$

and

$$\phi(n, z) \to z^{-n}, \quad \phi(n, z^{-1}) \to z^n \quad \text{for} \quad n \to -\infty,$$

respectively. Hence, their linear combinations

$$\varphi_+(n, z) = \alpha(z)\varphi(n, z^{-1}) + \beta(z)\varphi(n, z),$$
$$\varphi_-(n, z^{-1}) = \alpha(z^{-1})\varphi(n, z) + \beta(z^{-1})\varphi(n, z^{-1})$$

and

$$\phi_+(n, z) = \bar{\alpha}(z)\phi(n, z^{-1}) + \bar{\beta}(z)\phi(n, z),$$
$$\phi_-(n, z^{-1}) = \bar{\alpha}(z^{-1})\phi(n, z) + \bar{\beta}(z^{-1})\phi(n, z^{-1})$$

are determined by these complex numbers α, β and their conjugtes $\bar{\alpha}$, $\bar{\beta}$.

Assuming orthonormal relations among these asymptotic factors, there should be the following relations listed below:

$$\alpha(z)\bar{\alpha}(z^{-1}) + \beta(z)\bar{\beta}(z) = 1, \quad \alpha(z)\bar{\beta}(z^{-1}) + \beta(z)\bar{\alpha}(z) = 0,$$
$$\alpha(z^{-1})\bar{\alpha}(z) + \beta(z^{-1})\bar{\beta}(z) = 1, \quad \alpha(z^{-1})\bar{\beta}(z) + \beta(z^{-1})\bar{\alpha}(z^{-1}) = 0,$$
$$\bar{\alpha}(z)\alpha(z^{-1}) + \bar{\beta}(z)\beta(z) = 1, \quad \bar{\alpha}(z)\beta(z^{-1}) + \bar{\beta}(z)\alpha(z) = 0,$$
$$\bar{\alpha}(z^{-1})\alpha(z) + \bar{\beta}(z^{-1})\beta(z^{-1}) = 1, \quad \bar{\alpha}(z^{-1})\beta(z) + \bar{\beta}(z^{-1})\alpha(z^{-1}) = 0.$$

From these relations, we can derive

$$\bar{\alpha}(z) = \alpha(z), \quad \beta(z) = -\beta(z^{-1}), \quad \bar{\beta}(z^{-1}) = -\beta(z)$$

and

$$\alpha(z)\alpha(z^{-1}) = 1 + \beta(z)\beta(z^{-1}).$$

The last expression can be re-written as

$$|\alpha(z)|^2 = 1 + |\beta(z)|^2. \tag{8.11a}$$

Accordingly, to describe scatterings, it is convenient to define the function

$$S(n, z) = \frac{\varphi(n, z)}{\alpha(z)} = \varphi(n, z^{-1}) + R(z)\varphi(n, z) \quad \text{where} \quad R(z) = \frac{\beta(z)}{\alpha(z)}. \tag{8.11b}$$

The asymptotic form of $S(n, z)$ can then be written as

$$S(n, z) \to z^{-n} + R(z)z^n \quad \text{for} \quad n \to +\infty \tag{8.12c}$$

and

$$S(n, z) \to \frac{z^{-n}}{\alpha(z)} \quad \text{for} \quad n \to -\infty. \tag{8.12d}$$

To integrate the function $S(n, z)z^{m-1}$ in the complex plane, where $m \geqslant n$, along a closed circle C centered at the origin as illustrated by figure 8.4.

Writing that

$$\frac{1}{2\pi i}\oint_C \frac{\varphi(z)}{\alpha(z)} z^{m-1}\mathrm{d}z = \frac{1}{2\pi i}\oint_C \{\phi(n, z^{-1}) + R(z)\phi(n, z)\} z^{m-1}\mathrm{d}z \tag{8.13a}$$

are Cauchy's integrals. Using the relation

$$\phi(n, z) = \sum_{n'=n}^{\infty} K(n, n')\, z^n \tag{8.13b}$$

in the above integrals, the following expressions emerge from Cauchy's residues.

The first term in (8.13a) is

$$\frac{1}{2\pi i}\oint_C \phi(n, z^{-1})z^{m-1}\mathrm{d}z = \frac{1}{2\pi i}\sum_{n'=n}^{\infty} K(n, n')\oint_C z^{-n'+m+1}\mathrm{d}z = K(n, m) \quad \text{for} \quad m \geqslant n$$

and the second term

$$\frac{1}{2\pi i}\oint_C R(z)\phi(n, z)z^{m-1}\mathrm{d}z = \frac{1}{2\pi i}\sum_{n'=n}^{\infty} K(n, n')\oint_C R(z)z^{n'+m-1}\mathrm{d}z.$$

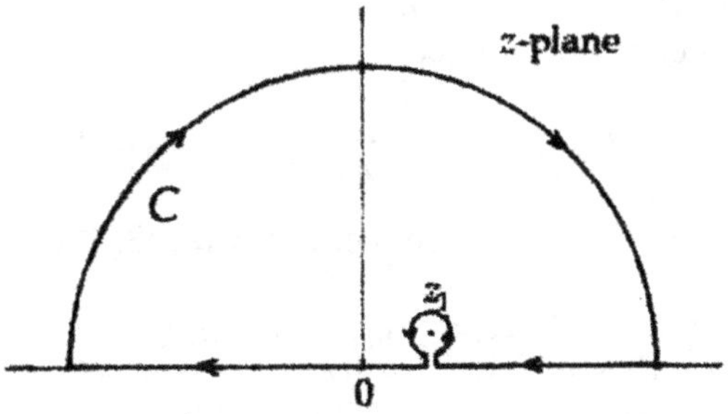

Figure 8.4. Cauchy's integral over a semi-circular path C, including a singular point $z = z_1$.

Therefore, defining $F(m) = \frac{1}{2\pi i}\oint_C R(z)\, z^{m-1}\mathrm{d}z$, the right side of (8.11a) can be expressed as

$$= K(n,\ m) + \sum_{n'=n}^{\infty} K(n,\ n')\, F(n' + m).$$

On the other hand, the left integral in (8.11a) can be evaluated as

$$\frac{1}{2\pi i}\oint_C \frac{\varphi(n,\ z)}{\alpha(n,\ z)} z^{m-1}\mathrm{d}z = I_\alpha + I_\mathrm{o} \tag{8.13c}$$

where I_α and I_o are residues at the poles of the integrand functions. The former arises from the properties of singular $z = z_j$ for $\alpha(z_j) = 0$, whereas the latter I_o is related to surfaces and domain boundaries for structural changes of crystals. In this case, the origin of I_o must be interpreted as internal adiabatic changes observed in thermal experiments.

In (8.13c), we have

$$\begin{aligned} I_\mathrm{o} &= -\sum_j \varphi(n,\ z) z^{m-1}\mathrm{Res}\left|\frac{1}{\alpha(z)}\right|_{z=z_j} = -\sum_j \phi(n,\ z_j)\, z_j^m c_j^2 \\ &= \sum_j c_j^2 \sum_{n'=n}^{\infty} K(n,\ n')\, z_j^{n'+m} \end{aligned} \tag{8.14a}$$

which is determined by the specific $z = z_j$. In addition, the singular properties of I_o can be evaluated by $\phi(n,\ z^{-1})$ in the vicinity of $z = 0$ in the function $S(n,\ z)$, satisfying the relation

$$a_{n-1}S(n-1,\ z) + a_nS(n+1,\ z) + b_nS(n,\ z) = \frac{z + z^{-1}}{2}S(n,\ z),$$

which can be approximated as

$$a_nS(n+1,\ z) \simeq \frac{1}{2z}S(n,\ z) \quad \text{for} \quad |z| \simeq 0.$$

Hence,

$$S(n,\ z) \simeq 2za_nS(n+1,\ z) \simeq \cdots \simeq (2n)^N a_na_{n+1}\ldots a_{n+N-1}S(n+N,\ z),$$

where $a_N = K(n+1,\ n+1)/2K(n,\ n)$.

Therefore, $S(n,\ z)$ can be evaluated by

$$S(n,\ z) \simeq \frac{1}{K(n,\ n)}z^{-n} \quad \text{at} \quad z \simeq 0.$$

Letting $-n + m - 1 = -1$ in the above, i.e. if $n = m$, (8.14a) can be expressed by

$$I_\mathrm{o} = \frac{1}{K(n,\ n)}\delta(n,\ m) \quad \text{for} \quad m \geqslant n \tag{8.14b}$$

where $\delta(n,\ m)$ is *Kronecker's delta.*

Writing $F = \sum c_j^2 z_j^n$ with regard to the pole of I_0, the left side of (8.13c) is

$$= \frac{\delta(n,\ m)}{K(n,\ n)} - \sum_{n'=n}^{\infty} K(n,\ n')\, F(n' + m),$$

which is equal to the right side of (8.13c)

$$= K(n,\ m) + \sum_{n'=n}^{\infty} K(n,\ n')\, F(n' + m).$$

Therefore, we have

$$\frac{\delta(n,\ m)}{K(n,\ n)} = K(n,\ m) + 2\sum_{n'=n}^{\infty} K(n,\ n')\, F(n' + m) \tag{8.14c}$$

where $F(m) = \frac{1}{2\pi i}\oint_C R(z) z^{m-1} \mathrm{d}z + \sum_j c_j^2 z_j^m$.

Equation (8.14c) is a GLM equation generalized for singularities of two poles of I_a and I_0. For $n \neq m$ in particular, (8.14c) is modified as

$$\frac{1}{K(n,\ n)^2} = 1 + F(2n) + \sum_{n'=n+1}^{\infty} \kappa(n,\ n')\, F(n' + m) \quad \text{where}$$
$$\kappa(n,\ m) = \frac{K(n,\ m)}{K(n,\ n)}. \tag{8.14d}$$

Using those GLM relations, the Toda lattice is said to be *analyzable or integrable.*

8.4 Entropy production from soliton potentials

The heat transfer process is the basic thermodynamic problem, depending on the detail of the environment in practice, which cannot be handled with respect to theories. Hence, the process can be discussed only in timescale τ of experimental practice; and often interpreted as related with temperature variation for entropy production.

We therefore discuss the energy transfer process with $S(n,\ z,\ \tau)$, considering (7.16b) for the equation $\tilde{\omega} = \frac{z - z^{-1}}{2i} = \sin q$ to be used as the factor $e^{-i\tilde{\omega}\tau}$.

Substituting the asymptotic expression for

$$S(n,\ z,\ \tau) = \{z^{-n} + R(n,\ \tau) z^n\} e^{-i\tilde{\omega}\tau} \tag{8.15a}$$

into the development equation

$$\frac{\mathrm{d}S(n,\ z,\ \tau)}{\mathrm{d}\tau} = a_{n-1} S(n-1,\ z,\ \tau) - a_n S(n+1,\ z,\ \tau),$$

we have

$$\begin{aligned}\frac{\mathrm{d}S(n,\ z,\ \tau)}{\mathrm{d}\tau} &\to \frac{1}{2}\left\{(z^{-n+1} - z^{-n-1}) + R(n,\ \tau)(z^{n-1} - z^{n+1})\right\}e^{-i\tilde{\omega}\tau} \\ &= \frac{z - z^{-1}}{2}\{z^{-n} - R(n,\ z)z^{n}\}e^{-i\tilde{\omega}\tau} \quad \text{for} \quad n \to +\infty.\end{aligned} \tag{8.15b}$$

Comparing (8.15*b*) with direct differentiation of (8.15*a*), we obtain

$$\frac{\mathrm{d}R(z,\ \tau)}{\mathrm{d}\tau} = \left(\frac{z - z^{-1}}{2} + i\tilde{\omega}\tau\right)R(z,\ \tau) = (z^{-1} - z)R(z,\ \tau),$$

leading to

$$R(z,\ \tau) = R(z,\ 0)e^{(z^{-1}-z)\tau}. \tag{8.15c}$$

Reference

[1] Gel'fand I M and Levitan B M 1955 *Am. Math. Soc. Transl.* **1** 253

IOP Publishing

Introduction to the Mathematical Physics of Nonlinear Waves (Second Edition)

Minoru Fujimoto

Chapter 9

Pseudopotentials and sine-Gordon equation: topological correlations in domain structure

The presence of domain structure in binary ordering systems is a well-known fact in nature, although the physical origin has not yet been understood. However, a possible mechanism appears to be attributed to the soliton theory, because of its basic dealing with the deformed lattice. Therefore, it should be logical to consider that the domain problem is likely related to an extended property from binary transitions, where the idea emerges with topological correlations for domain formation in crystals, as governed by the sine-Gordon equation.

Crystallographically, the finding of pseudopotentials in crystals were from microscopic studies with polarized light, demonstrating no relation with the symmetry group to constitute a domain boundary. From that viewpoint, it is interesting to note that the soliton theory is a new element of nonlinear physics, while the symmetry group of crystals characterizes the linear physics.

9.1 Pseudopotentials in mesoscopic phases

The space group in crystals are signified by translational symmetry, however modified if an additional rotational symmetry over *m*-times of unit lattice is constant. Illustrated in figure 9.1, such a *screw symmetry of m-fold rotation* creates potential energy $V_{\rm L}^{\rm m}(\theta_{\rm p})$ parallel to the order variable $\sigma_\perp(\theta_{\rm p})$ but in perpendicular directions to the m-fold screw axis. Therefore, we write it as expressed in the following.

As related to the order variables

$$\sigma_{\rm p} = \sigma_{\rm 0} e^{i\theta_{\rm p}}, \quad \text{where} \quad \theta_{\rm p} = \frac{2\pi}{\rm m}{\rm p} \quad \text{and} \quad {\rm p} = 0, 1, 2, \ldots, \ {\rm m} - 1, \tag{9.1a}$$

doi:10.1088/978-0-7503-3759-5ch9

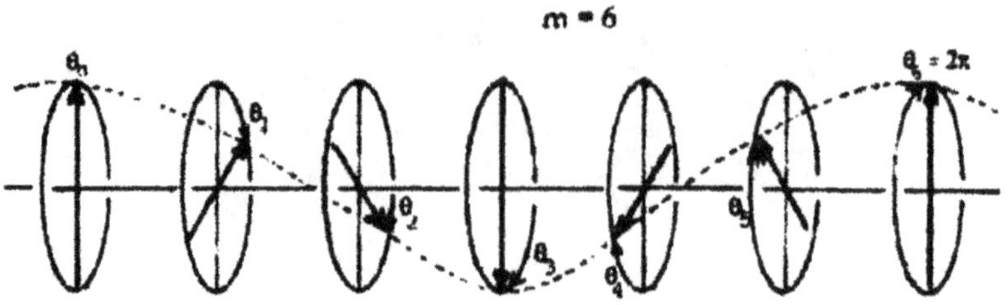

Figure 9.1. Phase variation of a collective pseudospin-mode pinned by a pseudopotential $V_m(\phi)$.

along the m-fold screw axis, we consider the potentials $V_m(\theta_p)||\sigma_p(\theta_p)$ expressed as

$$V_m(\theta_p) \propto \sigma_p = \sigma_o(e^{i\theta_p} + e^{-i\theta_p}) = 2\sigma_o \cos\theta_p. \tag{9.1b}$$

Here, angles θ_p are re-expressed by lattice coordinates $x_p = pa_o$ (a_o is a lattice constant), hence $\theta_p = \frac{2\pi}{m}\frac{x_p}{a_o} = G_m x_p$ where $G_m = \frac{2\pi}{ma_o}$. Accordingly, (9.1*b*) is re-written as

$$V_m(x_p) \propto \cos(G_m x_p). \tag{9.1c}$$

We notice in (9.1*c*) that $G_m x_p$ represents the Galilean phase, hence again replacing x_p by $m\phi$ for convenience of emphasizing the phase ϕ. Therefore, the potential of (9.1*b*) can be expressed as

$$V_m(\phi) = \frac{2\rho}{m}(\sigma^{im\phi} + \sigma^{-im\phi}) = \frac{2\rho\sigma_o^m}{m}\cos(m\phi), \tag{9.2}$$

where ρ is a proportionality constant, moving together with $\sigma_m(\phi)$ in finite amplitude, as specified by Weiss' law.

9.2 The sine-Gordon equation

The Weiss law is essential in nonlinear dynamics, indicating that $\sigma_m(\phi)$ is pinned by $V_m(\phi)$ in phase. It is realized that the former nonlinear variable is dominated by the phase ϕ, characterized by sufficiently large amplitude.

Therefore, we write the Gibbs function of $G(\sigma_m, \phi)$ to minimize for equilibrium, regarding the phase ϕ. First, expressing the Gibbs function as

$$G(\sigma) = \int_0^L \left\{\frac{a\sigma\cdot\sigma^*}{2} + \frac{b(\sigma\cdot\sigma^*)^2}{4} + \frac{mv_o^2}{2}\frac{\partial\sigma}{\partial x}\cdot\frac{\partial\sigma^*}{\partial x} + V_m(\phi)\right\}\frac{dx}{L},$$

which is modified as a time-average

$$\begin{aligned}\langle G(\sigma)\rangle &= G(\sigma_o, \phi)\\ &= \int_0^L \left\{\frac{a\sigma_o^2}{2} + \frac{b\sigma_o^4}{4} + \frac{mv_o^2}{2}\left(\frac{\partial\sigma_o}{\partial x}\right)^2 + \frac{mv_o^2\sigma_o^2}{2}\left(\frac{\partial\phi}{\partial x}\right)^2 + \frac{2\rho\sigma_o^m}{m}\cos(m\phi)\right\}\frac{dx}{L},\end{aligned}$$

where $\frac{\partial G(\sigma_0, \phi)}{\partial \sigma_0} = 0$ and $\frac{\partial G(\sigma_0, \phi)}{\partial \phi} = 0$ minimize $\langle G(\boldsymbol{\sigma})\rangle$, and obtain

$$a\sigma_0 + b\sigma_0^3 + 2\rho\sigma_0^{m-1}\cos(m\phi) + mv_0^2\left\{\sigma_0\left(\frac{\partial\phi}{\partial x}\right)^2 + \frac{\partial^2\sigma_0}{\partial x^2}\right\} = 0,$$

from which the time average can be expressed as

$$mv_0^2\sigma_0^2\frac{d^2\phi}{dx^2} - 2\rho\sigma_0^m\sin(m\phi) = 0 \quad \text{or} \quad \frac{1}{2}mv_0^2\sigma_0^2\left(\frac{d\phi}{dx}\right)^2 + V_m(\phi) = 0. \tag{9.3}$$

The second expression is realized as the energy conservation for $\boldsymbol{\sigma}(\phi)$, represented by the model of a simple pendulum of finite length in the gravitational field.

Using abbreviations $\psi = m\phi$ and $\xi = \frac{2m\rho\sigma_0^{m-2}}{mv_0^2}$, (9.3) can be simplified as

$$\frac{d^2\psi}{dx^2} - \xi\sin\psi = 0 \quad \text{and} \quad \frac{1}{2}\left(\frac{d\psi}{dx}\right)^2 - \xi\cos\psi = E, \tag{9.4}$$

where E is the energy of a pendulum. The first equation in (9.4) is known as the *sine-Gordon equation*, and the second one for energy conservation can be integrated as

$$x_1 - x_0 = \int_0^{\psi_1}\frac{d\psi}{\sqrt{2(E + \xi\cos\psi)}},$$

where the upper and lower limits ψ_1 and 0 correspond to x_1 and x_0, respectively.

This integral can be expressed by the *elliptic integral of the first kind*, if the modulus κ is defined by $\kappa^2 = \frac{2\xi}{E+\xi}$. Namely, if $\kappa^2 < 1$, defining $\frac{\psi}{2} = \theta$, we have

$$x_1 - x_0 = \frac{\kappa}{\sqrt{\xi}}\int_0^{\theta_1}\frac{d\theta}{\sqrt{1 - \kappa^2\sin^2\theta}},$$

where

$$\sin\theta_1 = \mathrm{sn}\frac{\sqrt{\xi}(x_1 - x_0)}{\kappa} \quad \text{for} \quad 0 < \kappa < 1 \quad \text{and}$$

$$\sin\theta_1 = \tanh\sqrt{\xi}(x_1 - x_0) \quad \text{for} \quad \kappa = 1.$$

For convenience, we define for $x_1 - x_0 = \Lambda(\kappa)$ to relate with $0 < \theta_1 < \pi$, and express

$$\Lambda(\kappa) = \frac{\kappa}{\sqrt{\xi}}\int_0^{\pi}\frac{d\theta}{\sqrt{1 - \kappa^2\sin^2\theta}} \tag{9.5}$$

as a periodic unit.

The pseudopotentials (9.2) are characterized by their phase $G_m x_p$ can now be specified by $\theta_p = \frac{2\pi}{m}p$, where the pseudospin waves $\psi(\theta)$ are perfectly pinned with no reflection from $V_m(\theta_p)$. However, noticed from the above discussion based on thermodynamic function $\langle G(\boldsymbol{\sigma})\rangle$, the pinning mechanism should be understood from the nonlinear dynamics of order variable $\boldsymbol{\sigma}$, which will be discussed in the next section.

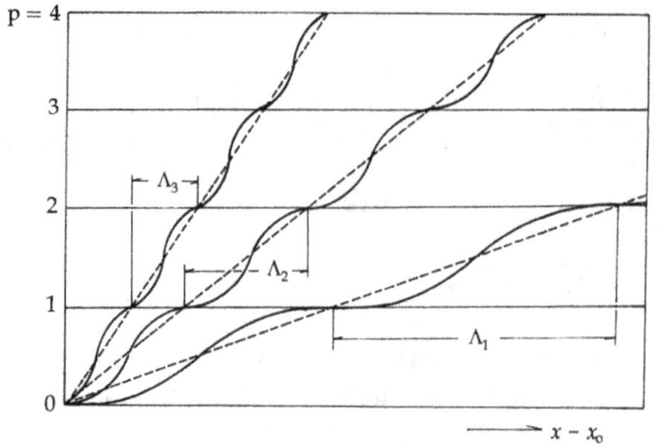

Figure 9.2. A phase function $\phi(x - x_0)$ in sine-Gordon equation for a pseudopotential $V_3(x - x_0)$. Reprinted with permission from [1], © 2010.

Nevertheless, the domain is characterized by phase inversion of $\sigma(\phi)$, i.e. $\phi_c \rightleftarrows -\phi_c$.

Figure 9.2 [1] illustrates the property of phase function $\psi(x - x_0)$, showing *nonlinear adiabatic variation* with respect to the parameter p.

9.3 Phase solitons in adiabatic processes

Nonlinear order variables in crystals are modeled by classical vector displacements, characterized by two components, which are substantiated by experiments. Also, their collective mode exhibits waves propagating along a well-defined direction described by a phase variable. Except for nodal portions, the wave pattern is dominated by the phase ψ, where the *transversal correlations* between propagating $\sigma(\psi)$. At the boundary wall, the order variable is $\sigma_\perp(\psi)$ in particular, where transversal correlations are significantly *topological.*

Considering a case where the longitudinal wave is predominantly a function of the phase, a domain boundary is specified by the perpendicular component $\sigma_\perp(\psi)$. We now assume that the correlations between $\sigma_{\perp 1}(\psi + \Delta\psi_1)$ and $\sigma_{\perp 2}(\psi + \Delta\psi_2)$ are signified by correlations between $\Delta\psi_1$ and $\Delta\psi_2$, attributing to the relation between pseudopotentials $V_{\perp 1}(\psi)$ and $V_{\perp 2}(\psi)$. As previously discussed, if considering Riccati's relation $\Delta\psi_2 = -\frac{1}{\Delta\psi_1}\frac{\mathrm{d}\Delta\psi_1}{\mathrm{d}y}$ along the coordinate y of transversal axis, those potentials compose Toda's lattice. Accordingly, it is logical to discuss the matter by means of the Klein–Gordon equation for the density $\rho_\perp = \sigma_\perp{}^*\cdot\sigma_\perp$ under the Weiss field proportional to $\rho_\perp$, i.e.

$$\frac{\partial^2\rho_\perp}{\partial\tau^2} - v^2\frac{\partial^2\rho_\perp}{\partial y^2} = -K(y, \tau)^2\rho_\perp. \tag{9.6a}$$

Replacing $K(y, \tau)$ by $\mathrm{m}(\psi)$ to deal with *convenience for singular properties* of ψ, (9.6*a*) is modified for the phase ψ as

$$\frac{\partial^2\psi}{\partial\tau^2} - v^2\frac{\partial^2\psi}{\partial y^2} + \mathrm{m}^2\psi = 0, \tag{9.6b}$$

for which a space–time transformation, $\xi = \frac{1}{2}(y - v\tau)$ and $\eta = \frac{1}{2}(y + v\tau)$, change (9.6*b*) to

$$\frac{\partial^2\psi}{\partial\xi\,\partial\eta} = \mathrm{m}^2\psi. \tag{9.6c}$$

Integrating (9.6*c*), we obtain for a critical case the following expressions:

$$\frac{\partial\psi}{\partial\xi} = \mathrm{m}^2\int\psi\mathrm{d}\eta + r(\xi) \quad \text{and} \quad \frac{\partial\psi}{\partial\eta} = \mathrm{m}^2\int\psi\mathrm{d}\xi + s(\eta),$$

where $r(\xi)$ and $s(\eta)$ are respectively arbitrary functions of ξ and η.

Further, we define $P = \frac{\partial\psi}{\partial\xi}$ and $Q = \frac{\partial\psi}{\partial\eta}$, and (9.6*c*) can be written as

$$Q\frac{\partial P}{\partial\psi} = P\frac{\partial Q}{\partial\psi} = \mathrm{m}^2\psi,$$

which can be integrated to obtain the relation $Q = a^2P$ with another constant a.

Therefore, we have the relations:

$$P(\psi) = \frac{\partial\psi}{\partial\xi} = \pm\frac{\mathrm{m}}{a}\psi \quad \text{and} \quad Q(\psi) = \frac{\partial\psi}{\partial\eta} = \pm\,(\mathrm{m}\;a)\psi, \tag{9.6d}$$

allowing for the phase function to be expressed as

$$\psi = e^{\pm i\mathrm{m}\zeta + b} = Ae^{\pm i\mathrm{m}(q\,y - \tilde{\omega}\,\tau)}, \tag{9.6e}$$

where

$$A = e^b, \quad \zeta = \frac{\xi}{a} + a\eta, \quad q = \frac{1 + a^2}{2a} \quad \text{and} \quad \tilde{\omega} = \frac{1 - a^2}{2a}.$$

Equation (9.6*e*) indicates that the phase reflects the m-fold symmetry of pseudopotential as exhibited by $\psi \sim e^{\pm i\mathrm{m}\zeta}$ where $\zeta = q\,y - \tilde{\omega}\,\tau$ and $\tilde{\omega} = vq$, showing the presence of lattice potential for *adiabatic processes* other than classical Weiss field.

9.4 The Bäcklund transformation and domain boundaries

It is noted that the binary inversion $\sigma_n(\boldsymbol{q}) \rightleftarrows -\sigma_n(\boldsymbol{q})$ at a lattice site n generates a critical splitting like quantum-mechanical splitting. In contrast, the domain inversion specified by $\boldsymbol{q} \rightleftarrows -\boldsymbol{q}$ is classical in nature, as illustrated in figure 3.2 in chapter 3, while accompanied by space–time fluctuations in magnitude of lattice parameters, consisting mathematically of a linear combination of domain fields. Therefore, in terms of wave functions, we generally consider a linear combination of ψ_1 and ψ_2 on both sides to deal with the domain structure, which is known as the *Bäcklund transformation*. It should be noticed in this case that the magnitude of finite vector $\boldsymbol{q}$

is *mesoscopic*, and the responsible fluctuations should be *topological* as related to correlations between domains.

We therefore consider that domain structure is signified by the presence of symmetric and antisymmetric fluctuations as defined by

$$\psi_A = \frac{\psi_1 + \psi_2}{\sqrt{2}} \quad \text{and} \quad \psi_P = \frac{\psi_1 - \psi_2}{\sqrt{2}}.$$

Then, assuming that ψ_A and ψ_P are described by either of Galilean phases ξ or η, we proceed to the same procedure as applied to (9.6*c*).

Namely, writing that

$$\frac{\partial(\psi_1 + \psi_2)}{\partial \xi} = \mathrm{m}^2 \int (\psi_1 + \psi_2)\, \mathrm{d}\eta + r(\xi) = P(\psi_1 + \psi_2)$$

and

$$\frac{\partial(\psi_1 - \psi_2)}{\partial \eta} = \mathrm{m}^2 \int (\psi_1 - \psi_2)\, \mathrm{d}\xi + s(\eta) = Q(\psi_1 - \psi_2),$$

which are considered for (9.6*d*) to be extended to two domains, 1 and 2.

Differentiating these relations, we obtain

$$\frac{\partial \psi_1}{\partial \eta}\frac{\partial P}{\partial \psi_1} + \frac{\partial \psi_2}{\partial \eta}\frac{\partial P}{\partial \psi_2} = \mathrm{m}^2(\psi_1 + \psi_2)$$

and

$$\frac{\partial \psi_1}{\partial \xi}\frac{\partial Q}{\partial \psi_1} - \frac{\partial \psi_2}{\partial \xi}\frac{\partial Q}{\partial \psi_2} = \mathrm{m}^2(\psi_1 - \psi_2).$$

Manipulating these derivatives, we arrange the above results in the following equations.

$$P\frac{\partial Q}{\partial \psi_1} - Q\frac{\partial P}{\partial \psi_1} = -2\mathrm{m}^2\psi_2, \quad P\frac{\partial Q}{\partial \psi_2} - Q\frac{\partial P}{\partial \psi_2} = 2\mathrm{m}^2\psi_2$$

$$\frac{\partial Q}{\partial \psi_1} - \frac{\partial Q}{\partial \psi_2} = 0 \quad \text{and} \quad \frac{\partial P}{\partial \psi_1} + \frac{\partial P}{\partial \psi_2} = 0.$$

Using these expressions for (9.6*c*), we obtain

$$P\frac{\partial^2 Q}{\partial \psi_A^2} - Q\frac{\partial^2 P}{\partial \psi_P^2} = 0,$$

hence, we have

$$\frac{\partial^2 P}{\partial \psi_P^2} + \kappa^2 P = 0 \quad \text{and} \quad \frac{\partial^2 Q}{\partial \psi_A^2} + \kappa^2 Q = 0, \tag{9.7a}$$

where $\kappa^2 = -\mathrm{m}^2$ or $\kappa = \pm\, i\mathrm{m}$ is arbitrary. Accordingly, we may write (9.6*d*) as

$$P(\psi_A) = \pm\frac{\mathrm{m}}{a}\psi_P \quad \text{and} \quad Q(\psi_P) = \pm(\mathrm{m}\, a)\psi_A, \tag{9.7b}$$

where

$$\psi_A = A(e^{i\mathrm{m}\zeta} + e^{-i\mathrm{m}\zeta}) = 2A\cos \mathrm{m}\zeta \quad \text{and} \quad \psi_P = A(e^{i\mathrm{m}\zeta} - e^{-i\mathrm{m}\zeta}) = 2A\sin \mathrm{m}\zeta. \tag{9.8a}$$

Therefore, we can express the wave equations for ψ_A and ψ_P as

$$\frac{\partial^2\psi_A}{\partial\tau^2} - v^2\frac{\partial^2\psi_A}{\partial y^2} = 0 \quad \text{and} \quad \frac{\partial^2\psi_P}{\partial\tau^2} - v^2\frac{\partial^2\psi_P}{\partial y^2} = -\sin\psi_P, \tag{9.8b}$$

implying that the phase modulation occurs in the phase mode ψ_P in (9.8*a*) governed by the *sine-Gordon equation* in (9.8*b*).

Nevertheless, paying attention to (9.7) specified by $\kappa = \pm\, i\mathrm{m}$, the domain boundary may not be composed of a single plane, but of two layers forming a wall of finite thickness, as implied by the solutions

$$P = p\sin\kappa\psi_P \quad \text{and} \quad Q = q\sin(\kappa\psi_A + \theta),$$

where θ is an arbitrary phase difference between P and Q. However, for convenience to obtain the relation (9.7*b*), we used the option $\theta = 0$ and $pq = -4$ for lattice modulation in each domain by the relations

$$\frac{\partial\psi_A}{\partial\xi} = \frac{2}{a}\sin\frac{\psi_P}{2} \quad \text{and} \quad \frac{\partial\psi_P}{\partial\eta} = \pm\, 2a\sin\frac{\psi_A}{2}. \tag{9.8c}$$

In this case, the lattice displacement vector in (9.6*a*) is dominated by the transversal component $\sigma_\perp(\zeta_0)$ proportional to $\psi_A(\zeta_0)$, and written as

$$\sigma_\perp(\zeta_0) \propto \tanh\left(\zeta_0 \pm \frac{\pi}{2}\right) = \pm\tanh\zeta_0. \tag{9.8d}$$

It is interesting that (9.8*c*) can be considered as the necessary requirement for linking domains as a single homogeneous system, which otherwise implies a possible physical extension across boundaries, as discussed in section 7.4 in chapter 7. In fact, there is an attempt to extend the concept of domain walls beyond Lax' theory, which will be discussed in chapter 10.

The lattice modulation outside the domain wall should be determined by the time-independent sine-Gordon equation (9.8*b*), i.e. the differential equation for a simple pendulum

$$\frac{d^2\psi_P}{d\zeta^2} = \pm\frac{1}{1 - v^2}\sin\psi_P,$$

providing solutions at closer points $\pm\zeta$ to boundary planes, which are expressed as

$$\psi_P(+\zeta) = 4\tan^{-1} e^{\frac{+\zeta}{\sqrt{1-v^2}}} \quad \text{and} \quad \psi_P(-\zeta) = 4\tan^{-1} e^{\frac{-\zeta}{\sqrt{1-v^2}}} \quad \text{for} \quad v^2 < 1.$$

It is clear from these solutions that we have an asymptotic relationship in the limit of $v \to \pm 1$; namely

$$\lim_{\zeta \to \zeta_0} \psi_P(\pm\zeta) \to \psi_P(\zeta_0) = \pm\frac{\pi}{2},$$

resulting in the following. That is

$$\psi_P(+\zeta_0) - \psi_P(-\zeta_0) = \Delta\psi_P(\zeta_0) = \pi, \tag{9.9a}$$

corresponding to the energy gap. Then, using (9.8*c*) for small ζ_0, we can replace $\psi_P(\zeta_0)$ by $\frac{d\psi_A}{d\zeta_0}$, (9.9*a*) is expressed as

$$\left\{\left(\frac{d\psi_A}{d\zeta}\right)_{+\zeta_0} - \left(\frac{d\psi_A}{d\zeta}\right)_{-\zeta_0}\right\}\Delta\zeta \propto \frac{\pi}{a}\Delta\zeta, \tag{9.9b}$$

where $\Delta\zeta = 2\zeta_0$, corresponding to *hypothetical work* for inversion $\psi_A(+\zeta_0) \to \psi_A(-\zeta_0)$ across the width $\Delta\zeta$ of domain wall. Writing this work as $\tilde{\sigma}\,\Delta\zeta$ for convenience, (9.9*b*) can be re-expressed as $\frac{d\ln\tilde{\sigma}}{d\tilde{\sigma}}$ = const., which is the same as Riccati's transformation [1]. Moreover, (9.9*b*) for a hypothetical work is interpreted as the imaginary potential $i\frac{d\psi_A}{d\zeta}$ discussed in chapter 3.

9.5 Computational studies of the Bäcklund transformation

The Bäcklund transformation provides a useful computational method to study multi-domain structure in thermal equilibrium. In this section, this method is discussed for numerical analysis, following Taniuchi and Nishihara [2].

We consider a system composed of multiple fields of domains. Each one can be signified by different timescales, while not necessarily required. They are driven by Klein–Gordon equations independently, but that is not necessary. Therefore, for the net field, we assume a symmetric linear combination as follows, expecting to describe stable configuration. Namely,

$$\psi = \sum_j \psi_j = \sum_j^n A_j e^{\pm im\zeta_j} \quad \text{where} \quad \zeta_j = \frac{\xi}{a_j} + a_j\eta \tag{9.10a}$$

and the constant a_j is arbitrary, depending however on the property of domain wall. Here, the number n indicates number of domains involved in the system, whereas A_j is related to factors associated with linear combination of inversion between correlated domains, defined as a significant parameter of transition as below.

$$A_j = \gamma_j A_{j-1} \quad \text{where} \quad \gamma_j = \frac{a_{j-1}c_A^2 - a_j c_P^2}{a_{j-1} - a_j}, \quad c_A^2 + c_P^2 = 1 \quad \text{for} \quad a_j \neq a_{j-1}. \tag{9.10b}$$

For performing transition from one domain field ψ_1 to another one ψ_3, Taniuchi and Nishihara carried out computer analysis via ψ_2, as illustrated in figure 9.3, which is

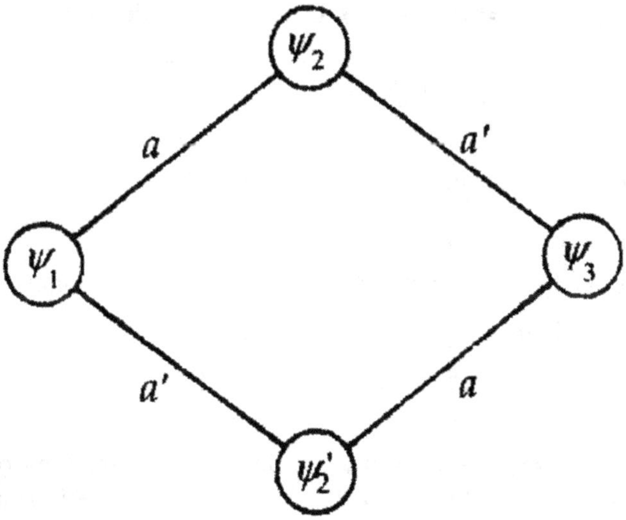

Figure 9.3. A computer simulation of phase solitons. Reproduced with permission from [2].

known as the *Lamb diagram*, where two values of the constant a are indicated as essential parameters for calculation.

Transitions were considered among domain fields

$$\psi_1 = A_1 e^{\pm im\zeta_1}, \quad \psi_2 = A_2 e^{\pm im\zeta_2} + \gamma_1 A_1 e^{\pm im\zeta_1} \quad \text{and}$$
$$\psi_3 = A_3 e^{\pm im\zeta_3} + \gamma_2 A_2 e^{\pm im\zeta_2} + \gamma_2\gamma_1 A_1 e^{\pm im\zeta_1},$$

however, we use the development operators, instead, to proceed to actual computation, as expressed by

$$\psi_2 = B_2\,\psi_1 \quad \text{and} \quad \psi'_2 = B'_2\,\psi_1,$$

thereby writing for paths from $\psi_1 \to \psi_2 \to \psi_3$ and $\psi_1 \to \psi'_2 \to \psi_3$ to express as

$$\psi_3 = B_3\psi_2 = B_3B_2\psi_1 \quad \text{and} \quad \psi_3 = B_3\psi'_2 = B_3B'_2\psi_1,$$

respectively, as illustrated in figure 9.3.

At any rate, since these paths specified by arbitrary parameters A_2 and A'_2 are adjustable mathematically for the unequal relation $B_3B_2 \neq B_3B'_2$ to be made equal in two different paths. These authors studied four hypothetical processes to be signified by two parameters a and a' in the Lamb diagram.

Using such a and a', the two paths are replaced by a single transformation $\psi_1 \to \psi_2 \to \psi_3$, for which we have a relation

$$\tan\frac{\psi_3}{4} = \frac{a+a'}{a-a'}\tan\frac{\psi_1 - \psi_2}{4},$$

from which we obtain

$$\psi_3 = 4\tan^{-1}\left\{\frac{a+a'}{a-a'}\,\frac{\sinh(\zeta_1 - \zeta_2)}{\sinh(\zeta_1 + \zeta_2)}\right\}. \tag{9.11}$$

Further, defining $\zeta_1 \pm \zeta_2 = \gamma_\pm(x - v_\pm \tau)$ for A and P modes, we have

$$v_\pm = \frac{1 + v_1 v_2 \mp \sqrt{(1 - v_1^2)(1 - v_2^2)}}{v_1 + v_2}.$$

We finally arrive at approximate expressions of ψ_3 from (9.11) as

$$\psi_3 \approx 4\tan^{-1}\{-e^{-\gamma_2(y - v_2\tau - \Delta_2)}\} \quad \text{for} \quad y > v_+,$$
$$\psi_3 \approx 4\tan^{-1}\{-e^{-\gamma_1(y - v_1\tau + \Delta_1)}\} \quad \text{for} \quad v_+ > y > v_- \quad \text{and}$$
$$\psi_3 \approx 4\tan^{-1}\{e^{\gamma_2(y - v_2\tau + \Delta_2)}\} \quad \text{for} \quad v_- > y, \quad \text{where}$$
$$\Delta_1 = \gamma_1 \ln\frac{a + a'}{a - a'} \quad \text{and} \quad \Delta_2 = \gamma_2 \ln\frac{a + a'}{a - a'}.$$

It is noted that these expressions should be interpreted in the limit of $\tau \to \pm\infty$ for the field ψ_3 to take values in the angular range of $0 \to \pm 2\pi \to 0$, respectively.

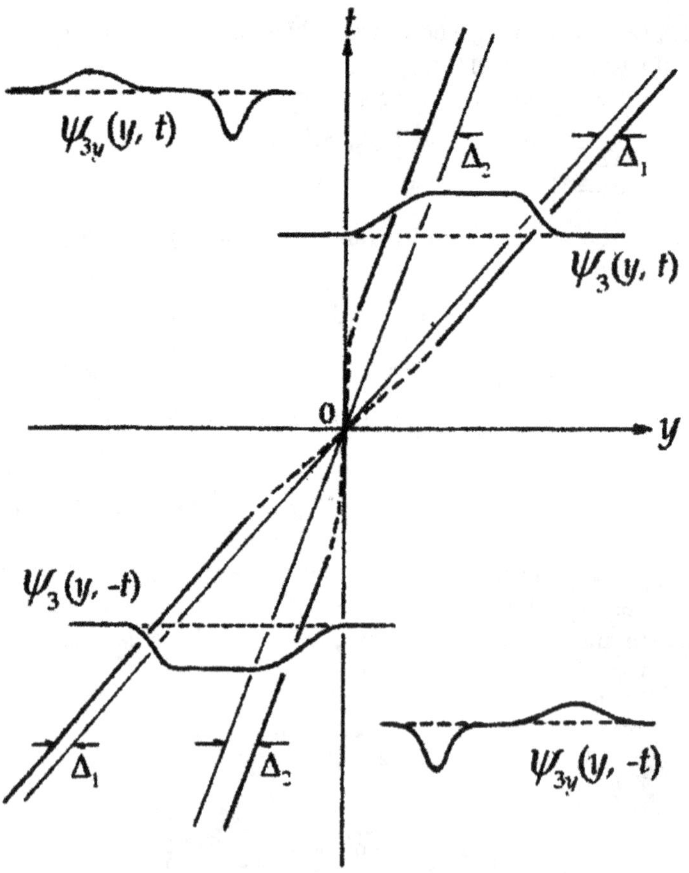

Figure 9.4. Reproduced with permission from [2].

Figure 9.4 is the reported result of numerical studies [2], showing that the *table-shaped curves* of calculated derivatives $\frac{\mathrm{d}\psi_3}{\mathrm{d}y}$ at a specified time τ, representing antisymmetric change of transversal correlations across a domain wall.

Exercise

1. In section 9.6, we set up the Klein–Gordon equation (9.6*a*) to modify as (9.6*b*), which was related to an adiabatic process in crystals. Analyzing in that way, discuss the complex potential $V_\perp(y) = \pm i\frac{\mathrm{d}u}{\mathrm{d}y}$ along the y-direction for dealing with inversion $\tanh y \rightleftarrows -\tanh y$ at the critical point.

References

[1] Fujimoto M 2010 *Thermodynamics of Crystalline States* (Berlin: Springer)
[2] Taniuchi T and Nishihara K 1998 *Nonlinear Waves* (Applied Math. Series in Japanese) (Tokyo: Iwanami)

Minoru Fujimoto

Chapter 10

Trigonal structural transitions: domain stability in topological order

Trigonal transitions have been observed in crystals, where pseudopotentials in C_3-cyclic symmetry is regarded as responsible. Most of these transitions were recognized earlier as incommensurate-to-commensurate transitions in charge-density waves, modulated lattice distortions, spin-density waves, spin-helical order, and surface layer structures. Among them we have typical examples of trigonal structure in crystal growing, as shown in figure 10.1, which is considered as the model of hexagonal patterns of domain structure, as observed optically from such transparent crystals. In this chapter, following Per Bak's article [1] about trigonal phase transition extensively, we discuss their significant features of *structural stability and topological order.*

10.1 The sine-Gordon equation

Shown in figure 10.1(a) shown a photographic image of a TSCC (tris-sarcosine calcium chloride) crystal under a polarizing microscope, where the presence of three differently oriented domains are clearly visible.

Symmetry axes are all identified from the direction of polarization, as indicated in figure 10.1(b), where the basic symmetry of order variables ψ_1, ψ_2 and ψ_3 are described by a phase variable ϕ in relation with this domain structure. In fact, these displacements occur independent of the structural change along each of the *b*-axes but slightly monoclinic *a*-axes. Therefore, on each *ab-plane*, these angular order variables are characterized by singularities across visible boundaries, as formulated by

$$\psi_{\rm m} = \psi_{\rm m+}e^{i\phi_{i+}} + \psi_{\rm m-}e^{i\phi_{i-}} \quad \text{and} \quad 0 \leqslant \phi_{\rm m\pm} \leqslant 2\pi, \tag{10.1}$$

where the phase of angular fluctuations are $\phi_1 = 0$, $\phi_2 = 60°$ and $\phi_3 = 120°$.

doi:10.1088/978-0-7503-3759-5ch10

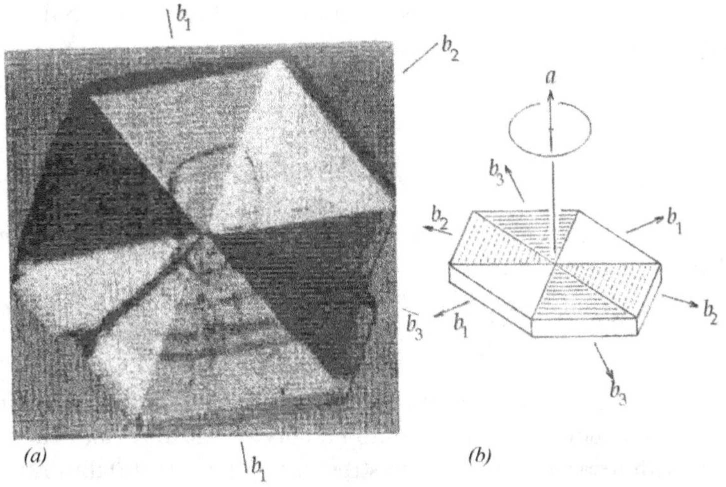

Figure 10.1. (*a*) A TSCC crystal photographed with a polarizing filter perpendicular to the *a*-axis. (*b*) Schematically shows the presence of vortex in the *bc*-plane, where b_1, b_2 and b_3 indicate directions of propagation in three trigonal domains.

Following McMillan [2], the phase correlation energy can be expressed for Gibbs' potential, which is then minimized to obtain the thermal equilibrium. According to him, the density Gibbs function Δg can be written for such a system in C_3-symmetry as follows.

$$\begin{aligned}\Delta g &= \sum_{\mathrm{m}}^{\mathrm{domain}} \Delta g_{\mathrm{m}} = \frac{1}{2}A\sum_{\mathrm{m}=1}^{3}\psi_m^2 + \frac{1}{4}B\sum_{\mathrm{m}=1}^{3}\psi_{\mathrm{m}}^4 + C(\psi_1\psi_2\psi_3 + \psi_{-1}\psi_{-2}\psi_{-3}) \\ &\quad + D\sum_{\mathrm{m}=1}^{3}\left(\psi_{\mathrm{m}}^3 + \psi_{\mathrm{m}}^{-3}\right) + E\sum_{\mathrm{m}=1}^{3}|(\nabla_{\mathrm{m}} - i\,\delta_{\mathrm{m}})\psi_{\mathrm{m}}| + F\sum_{\mathrm{m}=1}^{3}|(\boldsymbol{q}_{\mathrm{m}} \times \nabla_{\mathrm{m}})\psi_{\mathrm{m}}|,\end{aligned} \tag{10.2}$$

which is considered for phase correlations for the domain structure. Here (10.2) corresponds to wave vectors, $\boldsymbol{q}_{\mathrm{m}}$ are implicit, but related to the phase ϕ_{m}. The terms of *A*, *B*, *C* and *D* in (10.2) represent real response from the lattice and developing symmetry change. The terms of *D* compose a pseudopotential for C_3-symmetry. The last *E* and *F* terms compose complex potentials against lattice deformation. These terms are all additive, and responsible for the singular behavior of three boundaries m (1, 2, 3) under critical conditions.

It is realized that the terms of *E* and *F* need to be minimized to attain equilibrium, whereas all others are essential for the trigonal system in domain structure to stay in equilibrium. Nevertheless, calculating Δg for minimum, *E*-terms in these three domain boundaries, the derivative obtained as $\frac{\mathrm{d}\phi_{\mathrm{m}}}{\mathrm{d}y} - \delta$ represents essentially a field momentum, while the F-term gives additional potential. Accordingly, writing $\xi = \frac{D}{E}$ and $F\text{-term} = \sum_{\mathrm{m}} F_{\mathrm{m}}$, we obtain the following expression from $(\Delta g)_{\min}$, i.e.

$$\frac{1}{2}\left(\frac{\mathrm{d}\phi_{\mathrm{m}}}{\mathrm{d}y} - \delta\right)^2 - \xi\cos(3\phi_{\mathrm{m}}) - \frac{1}{2}\delta^2 + F_{\mathrm{m}} = 0, \tag{10.3a}$$

which is the dynamical equation for ϕ_{m} in the field of F_{m}. Differentiating this, we arrive at a sine-Gordon equation

$$\frac{\mathrm{d}^2\phi_{\mathrm{m}}}{\mathrm{d}y^2} - 3\xi\sin(3\phi_{\mathrm{m}}) = 0, \tag{10.3b}$$

which is considered for describing properties of trigonal phase transitions.

The significant achievement of the soliton theory is that the results can be analyzed geometrically with respect to the domain structure. Figure 10.2(*d*) illustrates the above arguments on domain stricture composed of two adjacent domains at a given temperature, as calculated using a simple pendulum model of vector order variable. Figures 10.2(*a*) and 10.2(*b*) sketch $\sigma_\perp(\phi)$ and $\sigma_\perp^2(\phi)$, showing singular peaks at domain boundaries ϕ_1, ϕ_2 and ϕ_3; characterized by discontinuity between $\pm\tanh\sigma_\perp$ in the former and $\mathrm{sech}^2\,\sigma_\perp$-peaks in the latter, respectively. Figure 10.2(*c*) shows that there is a critical limit for entropy production.

However, judging from (10.3*a*), there is generally ambiguity for the pseudopotential $F_{\mathrm{m}} - \frac{\delta^2}{2}$ between correlated domains. Considering angular functions $\psi_1(\phi)$, $\psi_2(\phi)$ and $\psi_3(\phi)$ of a circular lattice as characterized by soliton numbers $n-1$, n and $n+1$ in Toda's lattice, assuming $-\infty < \phi < +\infty$, the corresponding order variables should satisfy the wave equations

$$\frac{\mathrm{d}^2\sigma_{\perp n}}{\mathrm{d}\phi^2} + (\lambda_n - F_n)\sigma_{\perp n} = 0 \quad \text{and} \quad \frac{\mathrm{d}^2\sigma_{\perp n\pm 1}}{\mathrm{d}\phi^2} + (\lambda_{n\pm 1} - F_{n\pm 1})\sigma_{\perp n\pm 1} = 0.$$

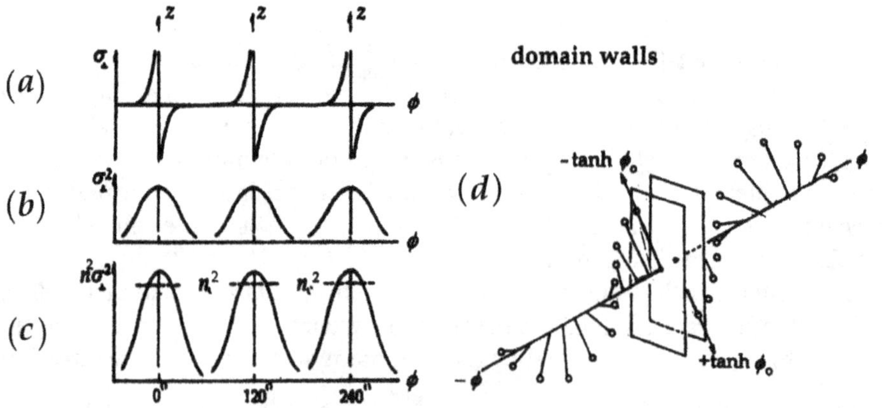

Figure 10.2. (*a*) Three kinks of phase solitons. (*b*) Corresponding weak peaks of three energy peaks in 120 degree difference. (*c*) Strong peaks over a critical soliton density n_c^2. (*d*) View of a domain wall.

Between unequal eigenvalues $\lambda_n \neq \lambda_{n\pm1}$, transitions $n \rightleftarrows n \pm 1$ are determined at domain boundaries by

$$\Delta\lambda_{n,n\pm1} = \Delta F_{n,n\pm1} = \mp 2n \ \text{sech}^2 \phi_c, \tag{10.3c}$$

which can be detected by thermodynamic experiments of adding and subtracting one soliton. As discussed in chapter 3, transitions (10.3*c*) can be measured in experiments under variable external pressure at a constant temperature.

10.2 Observing adiabatic fluctuations

At this point, it is interesting to recognize the LEED experiments carried out by Fain Jr and Chinn [3] on graphite surfaces under krypton gas pressure, exhibiting evidence for solitons in the pressurized structure. Illustrated in figure 10.3, the presence of solitons are evident in those parabolic curves of lattice distortion in soft-mode behavior, indicating that critical points depend on applied high pressure at different temperatures. In any case, their results support the concept of soliton that is consistent with the Landau theory.

Experimentally however, while inaccurate, it is convenient to consider for a uniaxial compression to replace uniform pressure, where the sample crystal can be of a flat plane in large area, so we may ignore volume change due to applied pressure. Thus, external force on a sample $-\boldsymbol{p} \cdot \boldsymbol{A}$ can be replaced by $-p\ A_\perp$ uniformly over the whole area $A_\perp$, where the counteracting external force to external pressure Δp should be described with the relation, $\Delta p = -\Delta p_{\text{int}}$.

Therefore, for a system specified internally by variable soliton number Δn, the Gibbs' function should change with Δn and Δp, as expressed by

$$\Delta G(\Delta n, \Delta p) \approx \frac{1}{2}a(\Delta n)^2 \mp A_\perp \Delta p \quad \text{for} \quad \Delta T = 0, \tag{10.4a}$$

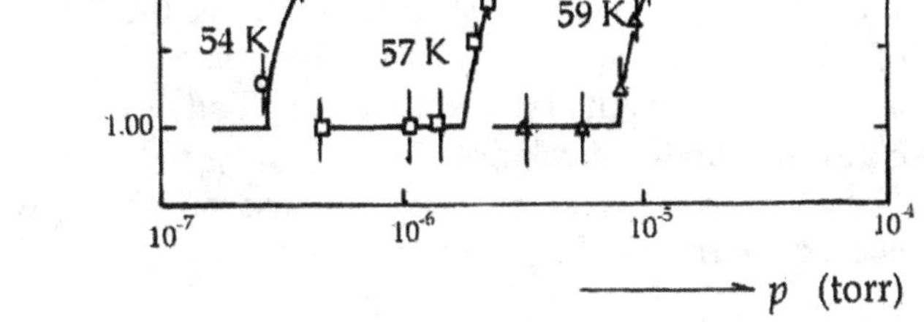

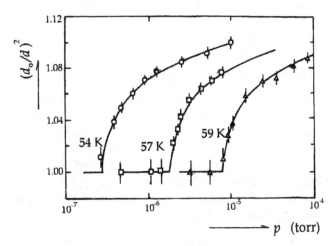

Figure 10.3. Adiabatic fluctuation process with varying pressure on graphite surface. Reproduced with permission from [3].

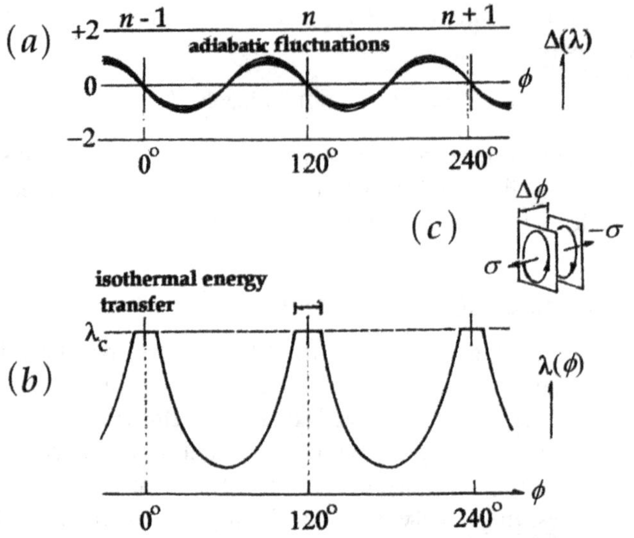

Figure 10.4. (*a*) Adiabatic fluctuation for $\Delta n = \pm 1$. (*b*) Isothermal energy transfers. (*c*) Inversion $\sigma \rightarrow -\sigma$ across the domain wall.

where a is a constant. The experimental results of Fain Jr and Chinn showed that

$$\Delta n \propto \sqrt{p_0 - p} \quad \text{where} \quad p_0 \text{ is a crtical pressure,} \tag{10.4b}$$

offering evidence of a soliton mechanism.

Shown in figure 10.4(*b*) are three $\text{sech}^2 \phi_m$ spectra where intensities are cut off for entropy production, and related adiabatic fluctuations for $\Delta n = 2$ are shown in figure 10.4(*a*). Also shown in figure 10.4(*c*) is that the inversion $\sigma_\perp \rightleftarrows -\sigma_\perp$ at domain boundaries is responsible for twist strains on the domain wall.

In addition, recent work on high-T_c superconductors revealed the soliton mechanism of the transition temperature T_c under variable pressure and related soft-mode behaviors, as discussed in chapter 11.

10.3 Toda's theory of domain stability

We have so far defined the domain wall mathematically, but in practice, they should be observed as separated domains in physical space. Therefore, Toda's approach is appropriate to a circularly correlated system in revealing the fluctuation in detail, since the path of correlations can be identified experimentally. In this section, we discuss the stability of domains in crystals, following Toda's discussion [4].

Assuming nearest-neighbor correlations among $\sigma_\perp(\psi_n)$ that is proportional to pseudopotential $V(\psi_n)$, we discuss the property of phase variable ψ_n with exponential correlations. Hence, we write the Lax wave equation as

$$\boldsymbol{L}\,\psi(n) = a_{n-1}\psi(n-1) + b_n\psi(n) + a_n\psi(n+1) = \lambda\psi(n), \tag{10.5}$$

where $-\infty < n < +\infty$. In such circularly infinite lattice, we have an eigen-equation

$$\frac{d^2\psi(n)}{d\phi^2} + \{\lambda_n - V_n(\phi)\}\psi(n) = 0 \tag{10.6a}$$

and

$$V_n(\phi) = V_n\left(\phi + \frac{2\pi}{N}\right). \tag{10.6b}$$

Here, the lattice may be converted to the other cyclic system, including C_2-symmetry, and N is any number, but 3 in the present case of a constant eigenvalue. Restricted to these circular-symmetries by exponential correlations, such correlations as C_3 and C_4 are forbidden, but may be correlated in other mechanisms in C_m, made possible by Bäcklund transformation.

Accordingly, in general, transformations are considered to carry on with a Wronskian matrix

$$\begin{pmatrix} \psi_1(N) \\ \psi_2(N) \end{pmatrix} = \boldsymbol{W}\begin{pmatrix} \psi_1(N+1) \\ \psi_2(N+1) \end{pmatrix} \quad \text{where} \quad \boldsymbol{W} = \begin{pmatrix} \psi_1(N) & \psi_1(N+1) \\ \psi_2(N) & \psi_2(N+1) \end{pmatrix}.$$

On the other hand, we write Lax' eigen-equations as

$$a_n\psi_1(n+1) + b_n\psi_1(n) + a_{n-1}\psi_1(n-1) = \lambda\psi_1(n),$$

and

$$a_n\psi_2(n+1) + b_n\psi_2(n) + a_{n-1}\psi_2(n-1) = \lambda\psi_2(n)$$

from which we obtain determinantal Wronskian as

$$W = a_n \begin{vmatrix} \psi_1(n) & \psi_1(n+1) \\ \psi_2(n) & \psi_2(n+1) \end{vmatrix} = a_{n-1} \begin{vmatrix} \psi_1(n-1) & \psi_1(n) \\ \psi_2(n-1) & \psi_2(n) \end{vmatrix}.$$

For rising $n \to N$ and descending $n \to 1$, we can write

$$W = a_N \begin{vmatrix} \psi_1(N) & \psi_1(N+1) \\ \psi_2(N) & \psi_2(N+1) \end{vmatrix} = a_o \begin{vmatrix} \psi_1(0) & \psi_1(1) \\ \psi_2(1) & \psi_2(0) \end{vmatrix} = a_o,$$

assuming that wavefunctions are normalized, i.e. $\begin{vmatrix} \psi_1(0) & \psi_1(1) \\ \psi_2(1) & \psi_2(0) \end{vmatrix} = 1$, if the initial conditions are given by

$$\psi_1(0) = \psi_2(1) = 1, \quad \psi_1(1) = \psi_2(0) = 0 \quad \text{and} \quad \Delta(\lambda) = \psi_1(0) + \psi_2(1) = 2,$$

thereby keeping the domain symmetry unchanged during transit to another. In this case, $a_N = a_o$, and we have

$$\det \boldsymbol{W} = \psi_1(N)\psi_2(N+1) - \psi_1(N+1)\psi_2(N) = 1.$$

On the other hand, the region between N and $N+1$, including the 3–4 case, we can write

$$c_1\psi_1(N) + c_2\psi_2(N) = \rho c_1\psi_1(0) = \rho c_1$$

and

$$c_1\psi_1(N+1) + c_2\psi_2(N+1) = \rho c_2\psi_2(1) = \rho c_2.$$

Therefore, ρ should satisfy the equation

$$\rho^2 - \Delta\lambda\,\rho + 1 = 0 \quad \text{where} \quad \Delta\lambda = \psi_1(N) + \psi_2(n+1) = \text{trace}\,\boldsymbol{W}.$$

Solving these equations, we obtain the relation

$$\rho = \frac{1}{2}\left\{\Delta\lambda \pm \sqrt{(\Delta\lambda)^2 - 4}\right\} \quad \text{where} \quad -2 < \Delta\lambda < +2. \tag{10.7a}$$

Hence, in terms of ρ, the range $-1 < \rho < +1$ determines a stable domain transition in C_3-symmetry.

Considering a case of static lattice for $a_{n\pm 1} = \frac{1}{2}$ and $b_n = 0$, (10.5) can be written as $\frac{1}{2}\{\psi(n-1) + \psi(n+1)\} = \lambda\psi(n)$, determining the stability limit as $(\Delta\lambda)^2 \geqslant 4$ from (10.7) or

$$\Delta\lambda_c - 2 = 0 \qquad \text{and} \qquad \Delta\lambda_c + 2 = 0 \tag{10.7b}$$

for the critical stability and opposite eigenvalue, respectively, both leading to entropy production, because of the real value of λ_c.

It is however significant in C_3-symmetry, if there are two different orientations of (ψ_1, ψ_2) on each of the facing boundary planes as related to topological correlations, expressing them by $\boldsymbol{u}_1$ and $\boldsymbol{u}_2$. Accordingly, we need to consider another Lax operator defined by $\boldsymbol{L}_{12} = \boldsymbol{u}_1{}^{-1}\boldsymbol{L}\,\boldsymbol{u}_2$, thereby writing this *auxiliary spectrum* as

$$\boldsymbol{L}_{12}\psi_1(N) = \lambda_1\psi_1(N) \quad \text{and} \quad \boldsymbol{L}_{12}\psi_2(N) = \lambda_2\psi_2(N), \tag{10.8}$$

we can derive the expression

$$(\lambda_1 - \lambda_2)\sum_{n=1}^{N}\psi_1(n)\psi_2(n) = a_N\{\psi_1(N+1)\psi_2(N) - \psi_1(N)\psi_2(N+1)\}$$
$$- a_0\{\psi_1(1)\psi_2(0) - \psi_1(0)\psi_2(1)\},$$

which does not vanish for nonzero N; hence we have $\lambda_1 \neq \lambda_2$.

In this context, the domain wall signified by the auxiliary condition can be at higher energies than stable conditions shown in figure 10.4(*b*). However, in the absence of $\boldsymbol{L}_{12}$, $\lambda_1 = \lambda_2 = \lambda_c$, implying that the domain structure may remain stable if there is no additional internal mechanism for such auxiliary operators. Using exponential correlations in the short range, it is notable that for C_2-symmetry, we have the relation $\Delta\lambda = \psi_0(1) + \psi_1(0) = 0$, indicating no adiabatic fluctuations. Between C_3 and C_4, we have $\Delta\phi = 45°$.

10.4 Kac's theory of nonlinearity for domain disorder

The mesoscopic domain wall is a phase transition zone between different orders, which can remain stable after the stress energy is removed to the elastic surrounding. Otherwise, the crystal is separated into independent domains. In the foregoing theory, domains are connected by Bäcklund's transformation, which is nonetheless mathematical convenience, but physically determined by Toda's potential.

However, considering that the correlations are essentially topological in terms of mesoscopic wavevector $\boldsymbol{q}$, so the auxiliary operator $\boldsymbol{L}_{12}$ and the corresponding $\boldsymbol{B}_{12}$ need to be modified to discuss the topological details.

We discuss the theory developed by Kac and his associate [5] based on Lax' basic equation (7.13*f*), as explained in the following. He proposed that (7.13*f*) is modified with the development operator as an additional term as described by $\Delta\boldsymbol{B}_{\mathrm{K}}$ for $\boldsymbol{B}_{12} = \boldsymbol{B}_{\pm} + \Delta\boldsymbol{B}_{\mathrm{K}}$ to satisfy the condition

$$\Delta\boldsymbol{B}_{\mathrm{K}}\boldsymbol{L} - \boldsymbol{L}\,\Delta\boldsymbol{B}_{\mathrm{K}} = 0. \tag{10.9a}$$

Namely, $\Delta\boldsymbol{B}_{\mathrm{K}}$ is assumed as in the type

$$\Delta\boldsymbol{B}_{\mathrm{K}} = \begin{pmatrix} 0 & \beta_1 & \gamma_1 & & \\ -\beta_1 & 0 & \beta_2 & \gamma_2 & \\ -\gamma_1 & -\beta_2 & 0 & & \\ & -\gamma_2 & & & \end{pmatrix} \tag{10.9b}$$

where nonvanishing terms $\pm\gamma_1, \pm\gamma_2, \ldots$ for $\Delta n = \pm 1$ are included to satisfy (10.9). It is noted that such $\Delta\boldsymbol{B}_{\mathrm{K}}$ is usable for some geometrical details involved in domain walls, hence we may assume that

$$\gamma_n = a_n\, a_{n+1} \quad \text{and} \quad \beta_n = (b_n + b_{n+1})a_n. \tag{10.9c}$$

In contrast, we assume that no such extension can be considered because of $\frac{\mathrm{d}\Delta L_{\mathrm{K}}}{\mathrm{d}\tau} = 0$.

Therefore, Kac's postulate is formulated as summarized below, i.e.

$$\Delta\boldsymbol{L}_{\mathrm{K}} = \begin{vmatrix} 0 & a_{n-1} & & \\ a_{n-1} & 0 & a_n & \\ & a_n & 0 & a_{n+1} \\ & & a_{n+1} & 0 \end{vmatrix},$$

$$\Delta\boldsymbol{B}_{\mathrm{K}} = \begin{vmatrix} 0 & 0 & a_{n-2}a_{n-1} & & \\ 0 & 0 & 0 & a_{n-1}a_n & \\ -a_{n-2}a_{n-1} & 0 & 0 & 0 & a_na_{n-1} \\ & -a_{n-1}a_n & 0 & 0 & 0 \\ & & -a_na_{n-1} & 0 & 0 \end{vmatrix}$$

and the developing equation is given by

$$\frac{\mathrm{d}\Delta \boldsymbol{L}_{\mathrm{K}}}{\mathrm{d}\tau} = \Delta \boldsymbol{B}_{\mathrm{K}}\Delta \boldsymbol{L}_{\mathrm{K}} - \Delta \boldsymbol{L}_{\mathrm{K}}\Delta \boldsymbol{B}_{\mathrm{K}}. \qquad (10.9d)$$

This can therefore be expressed as

$$\frac{\mathrm{d}a_n}{\mathrm{d}\tau} = a_n(a_{n+1}^2 - a_{n-1}^2). \qquad (10.9e)$$

Writing $a_n = \frac{1}{2}e^{-\frac{\phi_n}{2}}$ to define as τ instead of $\frac{\tau}{2}$, for convenience, (10.9e) can be converted to the equation of phase ϕ_n, i.e. $\frac{\mathrm{d}\phi_n}{\mathrm{d}\tau} = e^{-\phi_{n+1}} - e^{-\phi_{n-1}}$, allowing to select another phase in order for two independent phases to be transformable, like $\frac{\mathrm{d}\phi_{2n}}{\mathrm{d}\tau} = e^{-\phi_{2n-1}} - e^{-\phi_{2n+1}}$.

Accordingly, considering two new angular coordinates Q_n and Q_n' defined as

$$\phi_{2n} = Q_n' - Q_n, \quad \phi_{2n+1} = Q_{n+1} - Q_n' \quad \text{and} \quad \phi_{2n} + \phi_{2n+1} = Q_{n+1} - Q_n,$$

we have then

$$\frac{\mathrm{d}Q_n}{\mathrm{d}\tau} = e^{-\phi_{2n-1}} + e^{-\phi_{2n}} - \text{const.} \quad \text{and} \quad \frac{\mathrm{d}Q_n'}{\mathrm{d}\tau} = e^{-\phi_{2n}} + e^{-\phi_{2n+1}} - \text{const.}$$

and

$$\begin{aligned}\frac{\mathrm{d}^2Q_n}{\mathrm{d}\tau^2} &= -\dot{\phi}_{2n-1}e^{-\phi_{2n-1}} - \dot{\phi}_{2n}e^{-\phi_{2n}} \\ &= e^{-(\phi_{2n-2}+\phi_{2n-1})} - e^{-(\phi_{2n}+\phi_{2n+1})} = e^{-(Q_n-Q_{n-1})} - e^{-(Q_{n+1}-Q_n)}.\end{aligned} \qquad (10.10a)$$

Similarly, we have

$$\frac{\mathrm{d}^2Q_n'}{\mathrm{d}\tau^2} = e^{-(Q_n'-Q_{n-1}')} - e^{-(Q_{n+1}'-Q_n')}, \qquad (10.10b)$$

manifesting the presence of exponential potentials. Figure 10.5 illustrates the schematic view of Q- and Q'-lattices, showing there are correlations between these canonical modes for adiabatic transitions $\Delta n = \pm 1$ to be consistent with figure 10.4(b).

Reviewing the basic assumptions for Kac's model, transversal correlations are considered but the direction of propagation is disregarded; in fact the latter has no problem despite the 120° change in trigonal crystals. Therefore, the model may be applicable for other non-straight cases in general. Figure 10.6 shows the domain structure in C_3-symmetry, while figure 10.7 illustrates a simulated view of domain wall.

10.5 Domain separation and thermal and quasi-adiabatic transitions

Regarding the sine-Gordon equation (10.3b) for trigonal transition, the second term $\xi \cos 3\phi_{\mathrm{m}}$ on the right represents the pseudopotential determined by $V_3(\phi) = V_3(\phi + \frac{2\pi}{3})$.

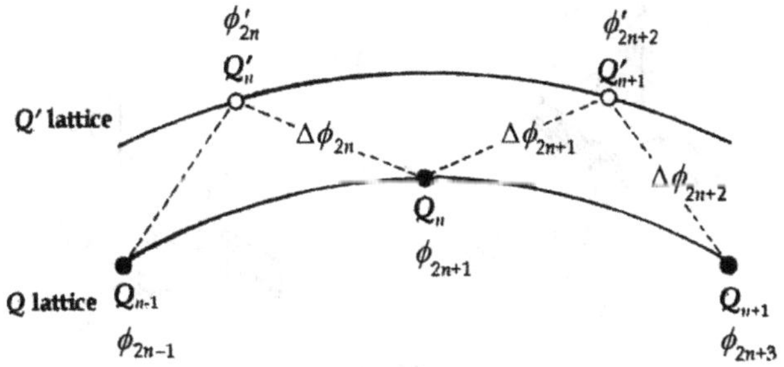

Figure 10.5. Schematic view of two Toda's lattices in a Kac's system, which are convertible by Bäcklund's transformation, showing that Q- and Q'-lattices are disoriented by rotations $\Delta\phi_{2n}$, $\Delta\phi_{2n+1}$ and $\Delta\phi_{2n-1}$ in the boundary zone.

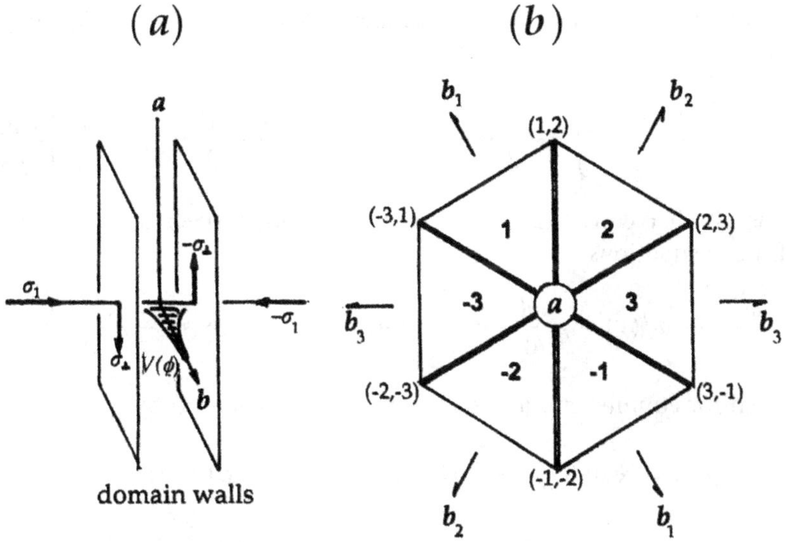

Figure 10.6. (a) Domain structure of a C_3 symmetry. On two boundaries, $\pm\sigma_\perp(\phi)$ exists before entropy production, which is converted to the potential $V_\perp(\phi)$ inside the wall. (b) Trigonal domains in hexagonal order in the bc-plane. Three domain walls are radially distorted but stable.

Therefore, for a sample crystal in single domain, the lattice potential can be represented by

$$V(\phi) = \frac{1}{2}\alpha\phi(y)^2 \quad \text{where} \quad -\frac{\pi}{6} < \phi(y) < \frac{\pi}{6}.$$

On the other hand, the emerging strain $\eta(y)$ should be considered to occur in the lattice, as coupled proportional to $\eta(y)\frac{\partial\phi}{\partial y}$, as inferred from the coupling of the discontinuity related to $\pm\sigma_\perp$ at $y = 0$ and $y = L$. Therefore, following Per Bak [1], we write the Gibbs free energy of the system composed of $\phi(y)$ and $\eta(y)$ as shown below.

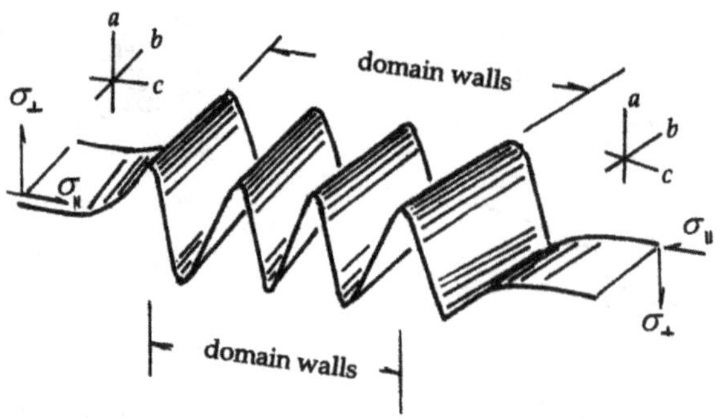

Figure 10.7. Simulated domain structure illustrated by $(\sigma_{\|}, \sigma_{\perp})$.

The Gibbs function density $G(y)$, defined from $G(T) = \int_{-L/12}^{+L/12} G(y)\mathrm{d}y$, can be written as

$$G(y) = \frac{1}{2}\left\{\frac{\mathrm{d}\phi(y)}{\mathrm{d}y} - \delta\right\}^2 + \frac{1}{2}\alpha\phi(y)^2 + \gamma\eta(y)\frac{\mathrm{d}\phi(y)}{\mathrm{d}y} + \frac{1}{2}\beta\eta(y)^2 + \frac{1}{2}\varepsilon\left\{\frac{\mathrm{d}\eta(y)}{\mathrm{d}y}\right\}^2$$

where ε is another constant. Minimizing $G(y)$ with respect to $\phi(y)$ and $\eta(y)$, we obtain Euler's equations

$$\frac{\mathrm{d}^2\phi(y)}{\mathrm{d}y^2} - \alpha\phi(y) + \gamma\frac{\mathrm{d}\eta(y)}{\mathrm{d}y} = 0 \quad \text{and} \quad \frac{\mathrm{d}^2\eta(y)}{\mathrm{d}y^2} - \beta\eta(y) - \gamma\frac{\mathrm{d}\phi(y)}{\mathrm{d}y} = 0$$

to be solved for equilibrium, and the solutions are expressed by

$$\phi(y) = A \sinh \kappa y, \; -\frac{L}{2} \leqslant y \leqslant \frac{L}{2}, \quad \phi(y + L) = \phi(L) + \frac{\pi}{3}$$

and

$$\eta(y) = B \cosh \kappa y, \quad \eta(y + L) = \eta(y)$$

where

$$\kappa^2 = \frac{1}{2\varepsilon}\left\{\beta + \alpha\varepsilon - \gamma^2 - \sqrt{(\beta + \alpha\varepsilon - \gamma^2)^2 - 4\alpha\varepsilon\beta}\right\},$$

$$A = \frac{\pi}{3}\sinh\frac{\kappa L}{2} \text{ and } B = \frac{A\gamma\kappa}{\varepsilon\kappa^2 - \beta}.$$

In this analysis, if the coupling γ is sufficiently strong in such a way that $(\sqrt{\alpha\varepsilon} + \gamma)^2 > \beta$, we have $\kappa = 0$, implying there is no steady solution for thermal processes. In that case, the domain structure is composed of two separate parts, keeping both $\phi(y)$ and $\eta(y)$ together in *topological fluctuations*. Otherwise, with $\kappa = 0$, we can continue to discuss these as separate systems, and the potential densities $\frac{1}{2}\alpha\phi\left(\frac{L}{2}\right)^2$ and $\frac{1}{2}\beta\eta\left(\frac{L}{2}\right)^2$ at the center $y = \frac{L}{2}$ are discussed as expressed between

$$\frac{\alpha A^2}{2} \quad \text{and} \quad \frac{\beta B^2}{2}\text{csch}^2\phi\left(\frac{L}{2}\right) = \frac{\beta B^2}{2}\left\{\coth^2\phi\left(\frac{L}{2}\right) - 1\right\},$$

and the energy difference should be signified by their thermal averages. Therefore, we consider the Boltzmann average of $\Delta E(\phi, \eta) = \frac{\beta B^2}{2}\coth^2\phi\left(\frac{L}{2}\right)$, which is written as

$$\langle\Delta E(\eta, \phi)\rangle = \frac{\beta B^2}{2}\tanh^2\phi\left(\frac{L}{2}\right) = \frac{\beta B^2}{2}\left\{1 - \text{sech}^2\phi\left(\frac{L}{2}\right)\right\}$$

and

$$\langle\Delta E(\phi, \eta)\rangle = -\frac{\beta B^2}{2}\tanh^2\phi\left(\frac{L}{2}\right) = \frac{\beta B^2}{2}\left\{\text{sech}^2\phi\left(\frac{L}{2}\right) + 1\right\},$$

so that we have the relations

$$\langle\Delta E(\phi, \eta)\rangle + \langle\Delta E(\eta, \phi)\rangle = 0 \quad \text{and} \quad \langle\Delta E(\phi, \eta)\rangle - \langle\Delta E(\eta, \phi)\rangle = \pm V_\perp\left(\frac{L}{2}\right),$$

where

$$\pm V_\perp\left(\frac{L}{2}\right) = \pm\frac{\beta B^2}{2}\text{sech}^2\phi\left(\frac{L}{2}\right), \tag{10.11}$$

showing the same soliton pulses proportional to $\pm\text{sech}^2\zeta$ for $\phi(\zeta)$ and $\eta(\zeta)$ with the phase $\zeta = k(y - vt)$ in the dynamical phase space.

Therefore, if the factor $\frac{\beta B^2}{2}$ representing the amplitude at $y = \frac{L}{2}$ is sufficiently larger than the critical n_c, the interaction energy $V_\perp\left(\frac{L}{2}\right)\|z$ can be transferred to the surroundings, confirming that *symmetrically modulated crystals* are consistent with thermodynamic equilibrium. Accordingly, we can consider the inversion $\left(\frac{d\psi_A}{d\zeta}\right)_\zeta \rightleftarrows -\left(\frac{d\psi_A}{d\zeta}\right)_{\zeta_0}$ along the z-axis, $2V_\perp\left(\frac{L}{2}\right)$ represents the hypothetical work for $\tanh\phi\left(\frac{L}{2}\right) \rightleftarrows -\tanh\phi\left(\frac{L}{2}\right)$.

10.6 Mesoscopic domains in topological disorder

In section 10.4, we discussed the hexagonal domain structure, paying no particular attention to topological order, which is however a significant natural problem. Basically, entropy production from forming domain walls is responsible for disordered structure, as commonly observed.

Further, significant findings are that the Weiss potential was considered to represent a real field of longitudinal lattice displacements, although imprecise, including non-reflective perpendicular potential, as familiar for electromagnetic cable transmission. Therefore, in thermodynamics, soliton potentials are dominated by sechϕ' and sech$^2\,\phi$, as shown already. There are some alternative expressions for the Korteweg–deVries equation, such as *the complex potentials* and *the related third-order Schrödinger equation,* which were however proposed for mathematical convenience to interpret the property of scatterers, but physically offering the same results from nonlinear dynamics.

The essential formula of the soliton potential is Bargmann's one-soliton formula (6.2) in chapter 6; i.e.

$$V(x,\, t) = -2\mu^2\ \mathrm{sech}^2(\mu x - 4\mu^3\tau), \tag{10.12}$$

which is given to rigid domains in thermodynamic equilibrium by assuming $\mu = \frac{\sqrt{v}}{2}$, where v is the speed of propagation in steady domains. Here, x and τ can represent *the phase of propagation* in crystals and *arbitrary timescale in topological domain movement for entropy production*, respectively.

For Korteweg–deVries' equation obtained in chapter 5, we considered the development operator of nonlinearity by $\mathcal{B} = a_1\mathcal{D} + a_3\mathcal{D}^3$ where $\mathcal{D} = \frac{\partial}{\partial x}$ and the term of $\mathcal{D}^2$ is absent. The reason for the latter is $(\mathcal{D}^2 - V)\psi = \varepsilon_2\psi$ and $\frac{d\varepsilon_2}{d\tau} = 0$ in the region of $V \to 0$, and hence $\frac{\partial\psi}{\partial\tau} = \mathcal{B}_3\psi = -4\mathcal{D}^3\psi$, after choosing $a_3 = -4$.

Therefore, with modulated amplitude $F(x, \tau) = F(x, 0)e^{ik^3\tau}$, such a wave function can be expressed as

$$\psi(x,\, \tau) = F(x,\, 0)e^{4ik^3\tau}(c_{21}e^{ikx} + c_{22}e^{-ikx}) \tag{10.13a}$$

where

$$c_{21}(k,\, \tau) = c_{21}(k,\, 0),\ c_{22}(k,\, \tau) = c_{22}(k,\, 0)e^{-4ik^3\tau}$$

and

$$m_{\mathrm{L}}(k,\, \tau) = -i\frac{c_{22}(k,\, \tau)}{c_{21}(k,\, \tau)} = m_{\mathrm{L}}(k,\, 0)e^{-8ik^3\tau}.$$

Hence, for the *non-reflective potential,* we consider a pole at an imaginary wave-vector $i\,k'' = k$, and can define a decaying reflection factor

$$R_{\mathrm{L}}(k'',\, \tau) = R_{\mathrm{L}}(k'',\, 0)e^{-8k''\tau}. \tag{10.13b}$$

It is noted that such a non-reflective potential exists as verified by a complex potential for a complex order variable, which is signified by the relation

$$V_{\perp}(x - vt,\, \tau) = V_{\perp}(x - vt)e^{-\gamma\tau} \quad \text{where} \quad \gamma = \alpha + i\,\beta. \tag{10.13c}$$

The modulated feature of a non-reflected potential can be expressed by assuming that

$$V_{\perp}(x,\, \tau) = Fe^{+i\phi} + F^{*}e^{-i\phi} \quad \text{and} \quad \phi = 2\alpha x + 8\alpha^3\tau. \tag{10.13d}$$

Substituting (10.13*d*) into Korteweg–deVries' equation with a common time τ, we can derive approximately the following expression, ignoring higher-order terms proportional to β^4, i.e. $|F|^2\frac{\partial F}{\partial x}$ and $\frac{\partial^3 F}{\partial x^3}$.

$$\frac{\partial F}{\partial \tau} - 12\alpha^2\frac{\partial F}{\partial x} - 6i\alpha\frac{\partial^2 F}{\partial x^2} - 12\,|F|^2 F = 0.$$

Transforming $x + 12\alpha^2\tau \to \bar{x}$ and $6\alpha\tau \to \bar{\tau}$, this equation becomes

$$i\frac{\partial F}{\partial \bar{\tau}} + \frac{\partial^2 F}{\partial \bar{x}^2} + 2\,|F|^2 F = 0 \tag{10.13e}$$

which is known as the *third-order Schrödinger equation*. In the space-time $(\bar{x}, \bar{\tau})$, the solution of (10.13*e*) is specified as

$$F(\bar{x}, \bar{\tau}) = 2\beta\, e^{4\beta^2\bar{\tau}}\,\text{sech}(2\beta\bar{x}), \tag{10.13f}$$

signifying boson fluctuations of $|F|^2$ in Kac's distorted domain structure sketched in figure 10.7.

Thus, it is now clear that Kac's theory can deal with disordered domain distribution, which is observed in practical systems, like trigonal arrangements.

However, unless precisely trigonal, the arrangement may be called topologically disorder, because the energy loss in unavoidable in domain walls.

Exercise

1. Using (10.4*a*), discuss that the domain structure can be simulated by repeated soliton waves.

References

[1] Bak P 1978 Solitons in incommensurate systems *Solitons in Condensed Matter Physics* ed A R Bishop and T Schneider (Berlin: Springer)

[2] McMillan W L 1976 *Phys. Rev.* B **14** 1496

[3] Fain S C and Chinn M D 1977 *J. Phys.* **38** C4–99
Fain S C and Chinn M D 1978 *Phys. Rev. Lett.* **39** 146

[4] Toda M 1997 *Nonlinear Lattice Dynamics* (*Applied Mathematics Series*) (Tokyo: Iwanami) (in Japanese) ch 4

[5] Kac M and Moerbecke P 1975 *Adv. Math* **16** 160

IOP Publishing

Introduction to the Mathematical Physics of Nonlinear Waves
(Second Edition)

Minoru Fujimoto

Chapter 11

Soliton theory of superconducting transitions

Recent experimental studies on superconductivity, including high-T_c superconductors in layer structure, have substantiated Fröhlich's modulated lattice, for which solitons are responsible. To initiate phase transitions, charged particles in a conduction band are responsible for normal and persistent currents, as described by London and Bardeen–Cooper–Schrieffer (BCS) theory. The modulated lattice arises with soliton excitation, forming Cooper pairs. With no other changes beyond theoretical predictions, the soliton mechanism is now regarded as essential for generating superconductivity [1]. Nevertheless, all theoretical predictions were detected in these materials, supporting all previous hypotheses for conducting crystals including cuprate- and iron-compounds in layer structures. In addition, recent experiments by Eremets and his associates on *metallic hydrogen-sulfide crystals* [2] have confirmed the presence of *proton currents*, exhibiting soft-mode behavior with respect to superconducting transitions at various temperatures. In this chapter, we discuss for the soliton theory to interpret superconducting phenomena in these conducting media.

11.1 The Meissner effect and Fröhlich's proposal

Superconductivity was first observed in anomalous electric resistance in metallic mercury, showing *zero-resistance* at temperatures 4.2 K and below. Meissner then discovered the diamagnetic superconductors, characterized by a structural change in the crystals at the transition temperature. All observed mechanisms so far substantiate the theories with respect to the modulated lattice structure.

Figures 11.1(*a*)–(*c*) illustrate the Meissner effect, where (*a*) and (*b*) show the magnetic induction $\boldsymbol{B}$ at temperatures above and at the critical point T_c, respectively, whereas (*c*) indicates the diamagnetization field $\boldsymbol{M}$ induced by the Meissner effect formulated as $\boldsymbol{B} = 0$, illustrated in figures 11.1(*b*) and (*c*) for a small test magnet to float in air at temperatures below T_c. In figure 11.2, the floating can be explained by

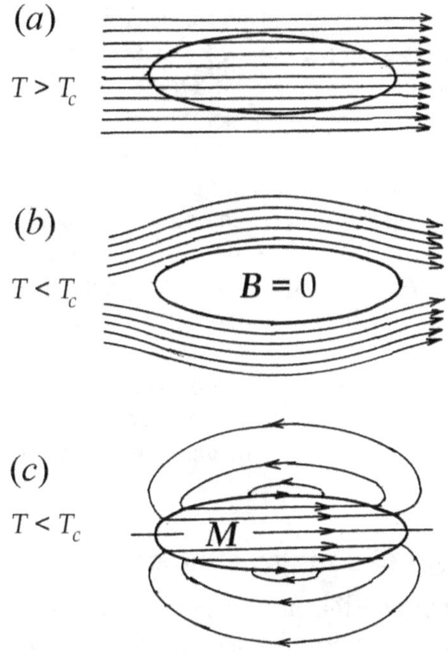

Figure 11.1. Meissner's effect. (*a*) Normal conductor for $T > T_c$ in an applied field $\boldsymbol{B} = \mu_o \boldsymbol{H}$. (*b*) Superconducting state for $T < T_c$, where all fluxes of $\boldsymbol{B}$ are pushed out, and inside is characterized by $\boldsymbol{B} = 0$. (*c*) Superconducting state for $T < T_c$ after removing $\boldsymbol{B}$. Inside is diamagnetic $\boldsymbol{M} = -\mu_o \boldsymbol{H}$.

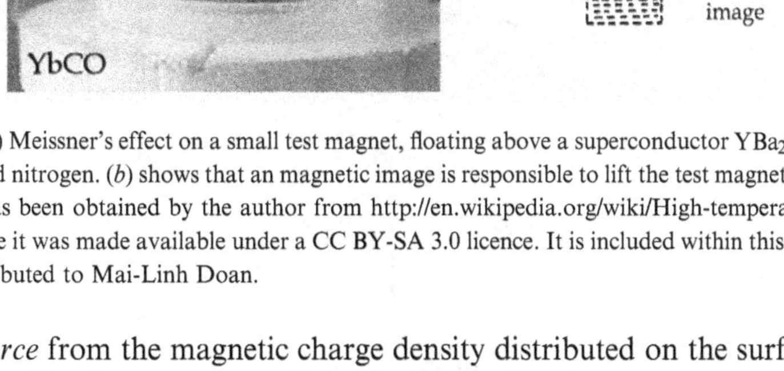

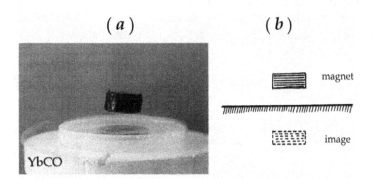

Figure 11.2. (*a*) Meissner's effect on a small test magnet, floating above a superconductor $YBa_2Cu_3O_7$(YBCO) cooled by liquid nitrogen. (*b*) shows that an magnetic image is responsible to lift the test magnet. This Meissner effect image has been obtained by the author from http://en.wikipedia.org/wiki/High-temperature_superconductivity, where it was made available under a CC BY-SA 3.0 licence. It is included within this chapter on this basis. It is attributed to Mai-Linh Doan.

an *image force* from the magnetic charge density distributed on the surface proportional to $e\,\Delta U = \nabla_r \cdot er$, where U is the lattice correlation energy and er is an induced dipole moment of displacement by charge attachment.

It is significant to consider that the Meissner effect took place in crystals in the presence of applied magnetic field, which was essential for the effect to be detected.

That can be understood with respect to the least-action principle, constituting the fact that the applied magnetic field $\boldsymbol{H}$ can be regarded as an *external thermodynamic quantity* for the internal magnetization $\boldsymbol{M}$. However, it is important that recent experimental results on high-T_c superconductors [3] support the concept of solitons in detail.

Figure 11.3 illustrates the crystal structure proposed for metallic hydrogen sulfide H3S by Eremets and his associates [2], showing the detail of hydrogen-bonding structure in H3S. From this illustration, we can assume that the protons 1 are in relatively weak hydrogen bonds, as can be displaced in a conduction band at applied high-pressures, being responsible for structural deformation due to protonic polarization.

In addition, table 11.1 [3] published in Wikipedia for cuprate-layer compounds indicate the data for variable transition temperatures to support the soliton mechanism, convincing for Fröhlich's mechanism to be responsible for superconductivity in general.

11.2 Magnetic images of Fröhlich's interaction

Experimentally, it is confirmed that the Meissner effect and zero-resistance represents the origin of superconducting phase transitions. Accordingly, it is logical to formulate it for the interaction potential between free charged particle and the modulated lattice.

We consider the interaction potential arising from $e\,\Delta U = e\,(\Delta \boldsymbol{r} \cdot \nabla_r)$, that constitutes the pinning potential of the *order variable*, as defined by Anderson in later discussion. Notably significant is that $e\,\Delta \boldsymbol{r}$ emerging at the critical point is a charged soliton vector that will modulate the lattice structure at a *specific wavevector* $\boldsymbol{q}$ *with the gradient* ∇_r; namely

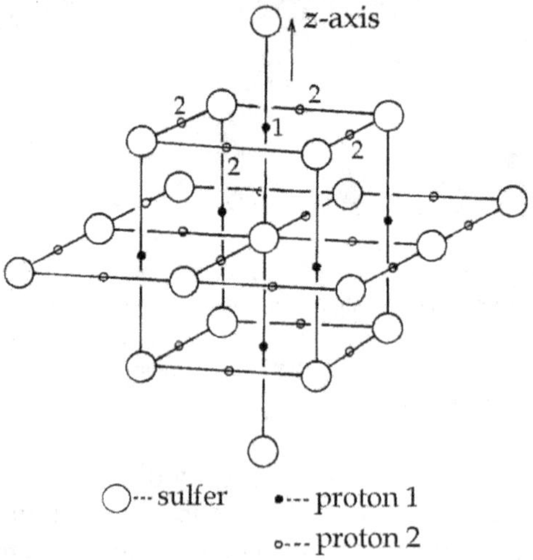

Figure 11.3. Lattice of the H3S phase in metallic hydrogen sulfide. At high pressure around 150 GPa along z-axis, protons 1 are displaced to be mobile by soliton potentials. Reproduced with permission from [8].

$$\Delta \boldsymbol{r} = \Delta \boldsymbol{r}_{\pm q} e^{i(\pm \boldsymbol{q} \cdot \boldsymbol{r} - \tilde{\omega}\tau)} \tag{11.1a}$$

in periodic structure. If the displacement occurs in favor of a symmetry axis along the z-axis in thermal equilibrium, (11.1a) can be given by

$$\Delta z = \Delta z_{\pm q} e^{i(\pm q\, z - \tilde{\omega}\tau)}, \tag{11.1b}$$

for which the soliton-excitation energy is expressed as $(\mathcal{H}_{\pm q})_{\rm o} = \lambda_{\rm o}\left(\tilde{b}^{\dagger}_{\pm \boldsymbol{q}}\tilde{b}_{\pm \boldsymbol{q}} + \frac{1}{2}\right)$ where $\tilde{b}^{\dagger}_{\pm \boldsymbol{q}}\tilde{b}_{\pm \boldsymbol{q}} = n_{\pm}$. And the soliton interaction is written as

$$\mathcal{H}_{\pm \boldsymbol{q}} = n_s \lambda_{\rm o} \tilde{b}^{\dagger}_{\pm \boldsymbol{q}}\tilde{b}_{\pm \boldsymbol{q}} + \text{const.} \quad \text{where} \quad n_s \lambda_{\rm o} = \lambda_{\boldsymbol{q}}. \tag{11.1c}$$

The above arguments on Fröhlich's interaction should then be subjected to the critical uncertainty, for which we need to discuss the Klein–Gordon equation on scattering of free charged particles by the density of displacement vector $\rho(\Delta z)$. Namely,

$$\left(\frac{\partial^2}{\partial \tau^2} - v^2\, \nabla^2_{z_{\rm o}}\right)\rho(\Delta z) = -K^2 \rho(\Delta z), \tag{11.2a}$$

where $K = -\frac{\Delta z}{v}$ represents the pseudopotential of a Weiss' field for the modulation, and $v = \tilde{\omega}/q$ is the velocity along the z-direction. Accordingly, such a modulation by (11.1c) should be determined by the thermal average

$$\langle \Delta z_{\pm q} | \int_{\Omega} \psi_{\boldsymbol{k}}(+\boldsymbol{q})\{\Delta z \cdot \nabla_{z_{\rm o}} V(z_{\rm o})\}\, \psi_{\boldsymbol{k}'}(-\boldsymbol{q}) {\rm d}\Omega | \Delta z_{\mp \boldsymbol{q}} \rangle_{\rm thermal}, \tag{11.2b}$$

which should be considered by a *canonical transformation* to minimize at thermal equilibrium. Here, $\langle \Delta z_{\pm \boldsymbol{q}}|$ and $|\Delta z_{\mp \boldsymbol{q}}\rangle$ are *bra* and *ket* of the displacements $\Delta z_{\pm \boldsymbol{q}}$ before and after charge scatterings, giving an unvarnished value of (11.2b) for $\boldsymbol{k}' - \boldsymbol{k} = \pm \boldsymbol{q} + \boldsymbol{G}$, but we can disregard $\boldsymbol{G} \neq 0$ for (11.2b).

Therefore, (11.1c) is modified for operators by the transformation (11.2b) as

$$\mathcal{H}_{\rm int} = \langle \mathcal{H}_{\pm \boldsymbol{q}} \rangle_{\rm thermal} = -e \int_{\Omega} {\rm d}^3 z \left\{\rho(z) \Delta z \cdot \nabla_{z_{\rm o}} V(z_{\rm o})\right\},$$

where $\rho(z) = \rho_{\boldsymbol{q}} \sum_n e^{-iqz_n}$ is the density of charged particles, and $\rho_{\boldsymbol{q}} = \sum_{\boldsymbol{k}} a^{\dagger}_{\boldsymbol{k}+\boldsymbol{q}} a_{\boldsymbol{k}}$. This expression can be written as

$$\mathcal{H}_{\rm int} = i\, e \sum_{\boldsymbol{q}} D_{\boldsymbol{q}}\left(\rho_{\boldsymbol{q}} \tilde{b}^{\dagger}_{\boldsymbol{q}} - \rho^{\dagger}_{\boldsymbol{q}} \tilde{b}_{\boldsymbol{q}}\right), \tag{11.2c}$$

where $D_{\boldsymbol{q}} \propto -\nabla_{z_{\rm o}} V(z_{\rm o})$, and $D_{\boldsymbol{q}}$ is related to the time-dependent phase $\phi_{\boldsymbol{q}}$ of modulation. Setting aside this time-dependence for convenience, the equation (11.2c) plays the essential role in the following discussion with respect to the inversion $+\boldsymbol{q} \rightleftarrows -\boldsymbol{q}$.

11.3 The Cooper pair and persistent current

It is significant that $\mathcal{H}_{\text{int}}$ in (11.2*c*) is related with inversion $+\boldsymbol{q} \rightleftarrows -\boldsymbol{q}$ for the following perturbation. Two free charged particles at wavevectors $\boldsymbol{k}$ and $\boldsymbol{k}'$ are involved in the Cooper pair [3], and we consider them to constitute the unperturbed Hamiltonian $\mathcal{H}_{\text{o}}$, and $\alpha\,\mathcal{H}_{\text{int}}$ as a perturbation. Here, the constant α should be selected for *conservative* perturbation, as determined by the canonical process.

With the Hamiltonian $\mathcal{H} = \mathcal{H}_{\text{o}} + \alpha\mathcal{H}_{\text{int}}$, we perform a canonical transformation defined by $\tilde{\mathcal{H}} = e^{-S}\mathcal{H}e^{S}$, where the function S should be selected to determine the value of constant α.

Expanding $\tilde{\mathcal{H}}$ for a weak value of S, we have

$$\tilde{\mathcal{H}} = \mathcal{H} + [\mathcal{H}, S] + \frac{1}{2}[[\mathcal{H}, S], S] + \cdots \simeq \mathcal{H}_{\text{o}} + \alpha\mathcal{H}_{\text{int}} + [\mathcal{H}_{\text{o}}, S] + [\alpha\mathcal{H}_{\text{int}}, S],$$

ignoring higher-order terms than S^2. If S is determined in such a way as

$$\alpha\mathcal{H}_{\text{int}} + [\mathcal{H}_{\text{o}}, S] = 0 \quad \text{and} \quad \tilde{\mathcal{H}} = \mathcal{H}_{\text{o}} + [\alpha\mathcal{H}_{\text{int}}, S], \tag{11.3a}$$

we confirm that $\alpha = eD_q$ for the canonical transformation.

Now that $\mathcal{H}_{\text{int}}$ is specified by eD_q and eigen values $\pm\lambda_q$, the perturbation is characterized by

$$\mathcal{H}_{\text{int}} = e\,D_q\begin{pmatrix} +\lambda_q & 0 \\ 0 & -\lambda_q \end{pmatrix} \tag{11.3b}$$

and the symmetric and antisymmetric eigenfunctions

$$\Delta z_{\text{A}} = \frac{\Delta z_{+q} + \Delta z_{-q}}{\sqrt{2}} \quad \text{and} \quad \Delta z_{\text{P}} = \frac{\Delta z_{+q} - \Delta z_{-q}}{\sqrt{2}}, \tag{11.3c}$$

corresponding to $\Delta n = 2$ and $\Delta n = 0$, respectively. Off-diagonal elements of S are given by

$$\langle \Delta z_{\text{A}}|S|\Delta z_{\text{P}}\rangle = -i\,e\,D_q\sum_k \frac{a^{\dagger}_{k-q}a_k}{\varepsilon_k - \varepsilon_{k-q} - \lambda_q} \quad \text{and} \quad \langle \Delta z_{\text{P}}|S|\Delta z_{\text{A}}\rangle = ieD_q\sum_{k'} \frac{a^{\dagger}_{k'+q}a_{k'}}{\varepsilon_{k'} - \varepsilon_{k'+q} + \lambda_q},$$

perturbing $\pm\lambda_q$ by $\begin{pmatrix} 0 & -eD_q \\ +eD_q & 0 \end{pmatrix}$. Accordingly, the resulting effect of $\mathcal{H}_{\text{int}}$ is determined by

$$\tilde{\mathcal{H}} - \mathcal{H}_{\text{o}} = \sum_q \frac{e^2 D_q{}^2}{2}\sum_{k,k'} a^{\dagger}_{k+q}a_k a^{\dagger}_{k'-q}a_{k'}\left(\frac{1}{\varepsilon_k - \varepsilon_{k-q} - \lambda_q} - \frac{1}{\varepsilon_{k'} - \varepsilon_{k'+q} + \lambda_q}\right). \tag{11.3d}$$

The off-diagonal elements in (11.3*d*) of the charge-density are associated with the coupling process, implying that two charged particles look chemically bonded. However, that is incorrect but only a statistical effect.

In figure 11.4(*a*), scatterings $\boldsymbol{k} \to \boldsymbol{k} - \boldsymbol{q}$ and $\boldsymbol{k}' + \boldsymbol{q} \to \boldsymbol{k}'$ cause *emission and absorption* of a soliton particle of energy λ_q between uniaxial displacements $+\Delta z_q$ and $-\Delta z_q$, respectively. Here, two terms in the brackets indicate singular behaviors of symmetrical solitons and antisymmetric particles at

$$|\Delta \boldsymbol{k}_c| = |\boldsymbol{q}_c| \text{ and } \Delta \boldsymbol{k}_c = \Delta \boldsymbol{k}_c' \text{ with } |\Delta z_{+q}|_c = |\Delta z_{-q}|_c \text{ at } T = T_c.$$

Therefore, we have relations

$$\Delta \boldsymbol{k}_c(1, 2) = -\Delta \boldsymbol{k}_c'(2, 1) \quad \text{and} \quad \boldsymbol{q}_c(1) \| \boldsymbol{q}_c(2).$$

In terms of the order variables $\boldsymbol{\sigma}_q(1)$ and $\boldsymbol{\sigma}_q(2)$ of charged particles, this parallel alignment is signified by $\boldsymbol{\sigma}_{q_c}(1) \cdot \boldsymbol{\sigma}_{q_c}(2) = \sigma_{q_c}^2$. In contrast, the antisymmetric relation between charged particles is attributed to Fermi's statistics.

Figure 11.4(*b*) illustrates the momentum space of two charged particles, where the conduction band is split into a circular gap $2\lambda_q$ at T_c due to two singularities in (11.3*d*). Summarizing two terms in (11.3*d*) to the critical temperature T_c as

$$\langle \tilde{\mathcal{H}} - \mathcal{H}_0 \rangle_{T_c} = e^2 D_q^2 \sum_{k_c, q_c} \frac{\lambda_q}{(\varepsilon_{k_c+q} - \varepsilon_{k_c})^2 - \lambda_{q_c}^2} a^\dagger_{k_c - q_c} a_{k_c} a^\dagger_{-k_c+q_c} a_{-k_c},$$

we have the potential of a *cluster of two complexes*

$$\mp V(2\boldsymbol{q}_c) = \mp e^2 D_{q_c}^2 \frac{\lambda_{q_c}}{(\varepsilon_{k_c+q_c} - \varepsilon_{k_c})^2 - \lambda_{q_c}^2} \tag{11.3e}$$

for a Cooper pair, having singularities signified by the relation

$$|\varepsilon_{k_c+q} - \varepsilon_{k_c}| \leqslant \lambda_{q_c}. \tag{11.3f}$$

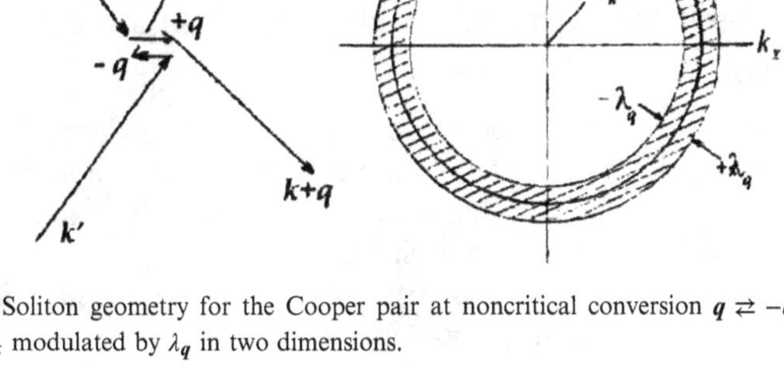

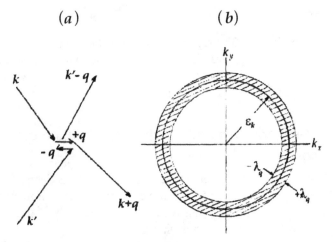

Figure 11.4. (*a*) Soliton geometry for the Cooper pair at noncritical conversion $\boldsymbol{q} \rightleftarrows -\boldsymbol{q}$. (*b*) One-particle energy surface ε_k modulated by λ_q in two dimensions.

The negative potential $-V(2\boldsymbol{q}_c)$ in (11.3c) is a pinning potential of *two order variables*, whereas $+V(2\boldsymbol{q}_c)$ is responsible for possible entropy production. Therefore, the former potential for pinning Cooper's pair $2e$ can be expressed as

$$-V(2\boldsymbol{q}_c) \propto e^2 D_{\boldsymbol{q}_c}^2 = e^2(n_{\mathrm{s,ini}}\lambda_{\boldsymbol{q}_c})^2 \propto -2u_0^2 \,\mathrm{sech}^2\, \phi_c, \tag{11.3g}$$

and the latter $+V(2\boldsymbol{q}_c)$ with the initial number of solitons $n_{\mathrm{s,ini}}$ can be for entropy production to the environment as

$$\mathrm{k_B}T_c = V(2\boldsymbol{q}_c) = e^2(n_{\mathrm{s,ini}}\lambda_{\boldsymbol{q}_c})^2 \quad \text{or} \quad T_c \propto e^2 n_{s,\mathrm{ini}}^2 \lambda_{\boldsymbol{q}_c}^2. \tag{11.3h}$$

It is significant to note that (11.3a) determines the critical temperature T_c for the Cooper pair of charges $2e$: i.e. proportional to $n_{s,\mathrm{ini}}^2$, exhibiting soft-mode behavior near T_c.

It is further fundamental to see that the soliton theory provides mobility to a Cooper pair to support the Meissner effect, responsible for *persistent current* arising from the super charges $2e$.

Noting that the complex factor $D_{\boldsymbol{q}_c}$ is expressed as $D'_{\boldsymbol{q}_c} e^{i\phi_c}$ is time-dependent because of time-dependent ϕ_c, (11.3f) can be written at T_c as

$$V(2\boldsymbol{q}_c) = |V(2\boldsymbol{q}_c)| e^{2i\phi_c}. \tag{11.4a}$$

Therefore, using a *continuity relation of charge-current*, we can obtain the *rotating persistent current density* from a time-derivative of (11.4a), i.e.

$$\boldsymbol{j}_\mathrm{s} = e\frac{\partial V(2\boldsymbol{q}_c)}{\partial \tau} = 2e|V(2\boldsymbol{q}_c)|\dot{\phi}_c\, e^{i(2\phi_c+\pi)} \tag{11.4b}$$

in the momentum space shown in figure 11.5(b), where $|V(2\boldsymbol{q}_c)|\dot{\phi}_c = v_\mathrm{s}$ represents the speed of Cooper's pair $2e$. Hence, we have relations

$$\boldsymbol{j}_\mathrm{s} = 2e\, v_\mathrm{s}, \quad m'\frac{\mathrm{d}\, v_\mathrm{s}}{\mathrm{d}\tau} = -2e\boldsymbol{E}_\mathrm{s} \quad \text{and} \quad \frac{\partial \boldsymbol{j}_\mathrm{s}}{\partial \tau} \propto \boldsymbol{E}_\mathrm{s}. \tag{11.4c}$$

Since $\boldsymbol{j}_\mathrm{s}$ of Cooper's pair $2e$ originates from displacements in the A-mode, the corresponding P-mode is compared in figure 11.5(a), showing no such currents. It is noted that the persistent current is characterized by $\mathrm{curl}\, \boldsymbol{j}_s \neq 0$; however in contrast, the normal Ohmic current is characterized by $\mathrm{curl}\, \boldsymbol{j}_\mathrm{n} = 0$.

11.4 Critical temperatures and energy gap in superconducting transitions

In section 11.3, we discussed the transition temperature T_c as determined by (11.3h) as well as the critical gap $2\lambda_q$ in (11.3f). However, the latter is not thermodynamic, whereas T_c represents equilibrium temperature. Accordingly, $2\lambda_q$ is only approximate, requiring corrections by a perturbing Cooper pair.

For a Cooper pair, the scattering processes $\pm\boldsymbol{k} \to \pm\boldsymbol{k}' \pm \boldsymbol{q}$ or $\Delta\boldsymbol{k} = \boldsymbol{k}' - \boldsymbol{k} \to \pm\boldsymbol{q}$ are in the center-of-mass system. Hence, in relative coordinates, the wavefunctions

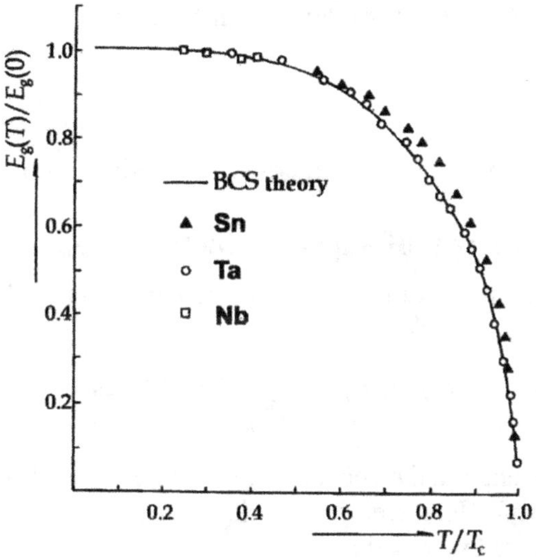

Figure 11.5. Energy gap ratio $E_g(T)/E_g(0)$ as a function T/T_c, comparing BCS theory with experimental results. Reproduced with permission from [1]. © 1986 John Wiley & Sons.

$\psi_{k',k}(z)$ of a Cooper pair should be specified by $z = 0$ at the center of coordinates. We therefore write a Schrödinger equation for $\psi_{k',k}(z) = \{\psi_{k'}(z), \psi_k(z)\}$ as

$$\left\{\frac{\hbar^2}{2m}\left(k^2 + \sum_{k'} k'^2\right) + V(2\boldsymbol{q}_c)\right\}\psi_{k',k}(z) = \varepsilon\, \psi_{k',k}(z),$$

where ε is an eigenvalue, and $\psi_{k',k}(z) = \sum_k \alpha_k e^{ik\cdot z}$, where α_k are for the linear combinations. Denoting the unperturbed energy by E_k, from the secular determinant, we derive the relation

$$(E_k - \varepsilon)\alpha_k + \sum_{k'} \langle \boldsymbol{k}, -\boldsymbol{k}| V(2\boldsymbol{q}_c) |\boldsymbol{k}', -\boldsymbol{k}'\rangle\, \alpha_{k'} = 0.$$

The perturbation $V(2\boldsymbol{q}_c)$ is distributed by $\alpha_k = \frac{C}{E_k - \varepsilon}$, and therefore

$$(E_k - \varepsilon)\alpha_k = -V(2\boldsymbol{q}_c)\int_{E_F}^{E_k} \alpha_{k'}\rho_{k'}\mathrm{d}E_{k'} \quad \text{where} \quad E_k - E_F = \Delta.$$

Here E_F is the Fermi energy in metals but represents equilibrium energies of the conduction bands in other conductors. Therefore, writing ρ_V in place of $\rho_{k'}$, we obtain

$$\frac{1}{\rho_V V(2\boldsymbol{q}_c)} = \int_{E_F}^{E_{k'}} \frac{\mathrm{d}E_{k'}}{E_{k'} - \varepsilon} = \ln\frac{E_{k'} - \varepsilon}{E_F - \varepsilon} = \ln\frac{E_{k'} - E_F + \Delta}{\Delta}.$$

Since we have $E_{k'} - E_F = \lambda_q$, this relation can be expressed as

$$\Delta = \frac{2\lambda_q}{e^{1/\rho_V V(2q_c)} - 1}, \tag{11.5}$$

indicating that the elemental soliton energy λ_q is predominant in the critical gap Δ.

11.5 Anderson's theory of superconducting phase transitions

In the field-theoretical approximation, following Anderson [4] the Hamiltonian for a Cooper pair can be written as

$$\mathcal{H} = \sum_k (\varepsilon_k a_k^\dagger a_k + \varepsilon_{-k} a_{-k}^\dagger a_{-k}) - \sum_{k,-k} V(\boldsymbol{k}, -\boldsymbol{k}; \boldsymbol{2q}_c) a_k^\dagger a_{-k}^\dagger a_k a_{-k}, \tag{11.6a}$$

where $\varepsilon_k = \varepsilon_{-k}$ represent equal eigenvalues of a single charged particle at $\boldsymbol{k}$ and $-\boldsymbol{k}$ in the conduction band. Here, free one-particle energy is assumed degenerate at $\varepsilon_{\pm k}$, hence (11.6a) can be re-expressed as

$$\mathcal{H} = \sum_{k,k'} \varepsilon_k a_k^\dagger a_k - \sum_{k,k';q_c} V(\boldsymbol{k}, \boldsymbol{k}'; 2\boldsymbol{q}_c) a_{k'+q_c}^\dagger a_{k-q_c}^\dagger a_{k'} a_k. \tag{11.6b}$$

On the other hand, the wavefunction can be written for soliton numbers n_k, $n_{k'}$ and the like, taking values either 1 or 0 in $\psi(\ldots, n_k, \ldots, n_{k'}, \ldots)$, signifying these states are occupied or not, respectively. In BCS theory, in contrast, the Cooper pair was described by a two-particle wavefunction $\psi(n_k, n_{k'})$, where $n_k = a_k^\dagger a_k$ and $n_{-k} = a_{-k}^\dagger a_{-k}$, hence, we use the BCS Hamiltonian for Cooper pairs, i.e.

$$\mathcal{H}_{\mathrm{BCS}} = -\sum_k (1 - n_k - n_{-k})\, \varepsilon_k - V(\boldsymbol{k}, -\boldsymbol{k}; 2\boldsymbol{q}_c) \sum_{k,-k} a_k^\dagger a_{-k}^\dagger a_{-k} a_k. \tag{11.6c}$$

In (11.6c), we therefore define for each one-particle state to be occupied or unoccupied by a Cooper pair, as expressed below.

$$(1 - n_k - n_{-k})\psi(1_k, 1_{-k}) = -\psi(1_k, 1_{-k}) \quad \text{and} \quad (1 - n_k - n_{-k})\psi(0_k, 0_{-k}) = \psi(0_k, 0_{-k}),$$

where these wavefunctions can be rewritten in the matrix form as $\begin{pmatrix}0\\1\end{pmatrix}$ and $\begin{pmatrix}1\\0\end{pmatrix}$, respectively; and

$$1 - n_k - n_{-k} = \begin{pmatrix}1 & 0\\ 0 & -1\end{pmatrix} = \sigma_k(z)$$

which is similar to corresponding z-component of a Pauli's matrix $\boldsymbol{\sigma}$. With this matrix formulation, we can write that

$$a_k^\dagger a_{-k}^\dagger \psi(1_k, 1_{-k}) = 0 \quad \text{and} \quad a_k^\dagger a_{-k}^\dagger \psi(0_k, 0_{-k}) = \psi(1_k, 1_{-k}),$$

where the operator $a_k^\dagger a_{-k}^\dagger$ is assigned to the x- and y-components of Pauli's spin matrix,

$$\sigma_k(x) = \begin{pmatrix} 0 & 1 \\ 1 & 0 \end{pmatrix} \quad \text{and} \quad \sigma_k(y) = \begin{pmatrix} 0 & -i \\ i & 0 \end{pmatrix},$$

and

$$\sigma_k^+ = \sigma_k(x) + i\,\sigma_k(y) = \begin{pmatrix} 0 & 2 \\ 0 & 0 \end{pmatrix} \quad \text{and} \quad \sigma_k^- = \sigma_k(x) - i\,\sigma_k(y) = \begin{pmatrix} 0 & 0 \\ 2 & 0 \end{pmatrix}.$$

Thus, we obtain the relations

$$a_k^\dagger a_{-k}^\dagger = \frac{1}{2}\sigma_k^- \quad \text{and} \quad a_{-k}a_k = \frac{1}{2}\sigma_k^+, \tag{11.6d}$$

representing the density matrix of a Cooper pair, constituting Anderson's theory of superconducting phase transitions.

Using Anderson's order variable vector $(\sigma_k(x), \sigma_k(y), \sigma_k(z))$, the thermal Hamiltonian can be formulated as

$$\begin{aligned} \langle \mathcal{H}_{\text{BCS}} \rangle_{\text{thermal}} &= -\sum_k \varepsilon_k \sigma_k(z) - \frac{V(2\boldsymbol{q}_c)}{4} \sum_{k',k} \sigma_{k'}^- \sigma_k^+ \\ &= -\sum_k \varepsilon_k \sigma_k(z) - \frac{V(2\boldsymbol{q}_c)}{4} \sum_{k',k} \{\sigma_{k'}(x)\sigma_k(y) + \sigma_{k'}(y)\sigma_k(x)\} \\ &= -\sum_k \boldsymbol{\sigma}_k \cdot \boldsymbol{F}_k \quad \text{where} \end{aligned} \tag{11.6e}$$

$$\boldsymbol{F}_k = \left(-\frac{V(2\boldsymbol{q}_c)}{2} \sum_k \sigma_k(x),\ -\frac{V(2\boldsymbol{q}_c)}{2} \sum_k \sigma_k(y),\ \varepsilon_k \right),$$

representing the *Weiss field energy* in crystals.

Also, accepting the fact of the order variable as a classical vector, we can analyze the critical gap geometrically as follows. Considering (11.6*e*) in the xy-plane that $(\boldsymbol{\sigma}_k \| \boldsymbol{F}_k)_{xy}$, we define the angle θ_k between $F_k(x)$ and $F_k(z)$ as

$$\tan\theta_k = \frac{F_k(x)}{F_k(z)} = \frac{\sigma_k(x)}{\sigma_k(z)} = \frac{V_{k'}(2\boldsymbol{q}_c)}{2\varepsilon_k}\sin\theta_k \quad \text{where} \quad V_{k'}(2\boldsymbol{q}_c) = \sum_{k'} V(\boldsymbol{k}', \boldsymbol{k}; 2\boldsymbol{q}_c)\,\sigma_{k'}.$$

Such a phase angle as θ_k is associated with mesoscopic modulation by a superconducting cluster, however we pay attention to singularity in the equation $\tan\theta_k = \frac{V_{k'}}{2\varepsilon_k}$ at $\theta_k = 0$, as related to the distributed $V_{k'}$, which we must accept for the process of canonical transformation.

Therefore, writing this relation as $\Delta_k = \sum_{k'} \frac{V_{k'}}{2} \frac{\Delta_k}{\sqrt{\Delta_k^2 + \varepsilon_{k'}^2}}$ i.e. $1 = \frac{V_k}{2} \sum_{k'} \frac{1}{\sqrt{\Delta_k^2 + \varepsilon_{k'}^2}}$, replacing $\sum_{k'} \ldots$ by an integral over distributed $-\lambda_q < \varepsilon_k < +\lambda_q$, this equation can be re-expressed by $1 = \frac{V_k \rho_F}{2} \int_{-\lambda_q}^{+\lambda_q} \frac{d\varepsilon}{\sqrt{\Delta^2 + \varepsilon^2}} = V_k\,\rho_F \sinh^{-1}\frac{\lambda_q}{\Delta}$, leading to the equation

$$\Delta = \frac{\lambda_{q_c}}{\sinh \frac{1}{V_k \rho_F}} \simeq 2\lambda_{q_c} e^{-\frac{1}{V_k \rho_F}} \quad \text{for} \quad V_k > 0 \tag{11.6f}$$

which is mathematically almost identical to (11.5) and dominated by the relation with modulation amplitude $2\lambda_{q_c}$.

Figure 11.5 shows experimental results of observed critical gaps in representative metals, which was compared with the BCS formula, indicating reasonable agreement.

11.6 Cuprate-layer structure and the Cooper pair

The structure of the conducting phase in H3S shown in figure 11.2 is relatively clear to understand Fröhlich's polar attachment of free charged particles. Although more complicated, the crystallographic model for interfacing cupric layers shown in figure 11.6 indicates the presence of polar attachment across two cupric layers, where unpaired electrons in 3d($x^2 - y^2$) orbitals are eminent as rotatable in layer structure. A similar situation for unpaired 3d electrons in the layer structure of Fe^{3+}-compounds exists, but has not yet been investigated.

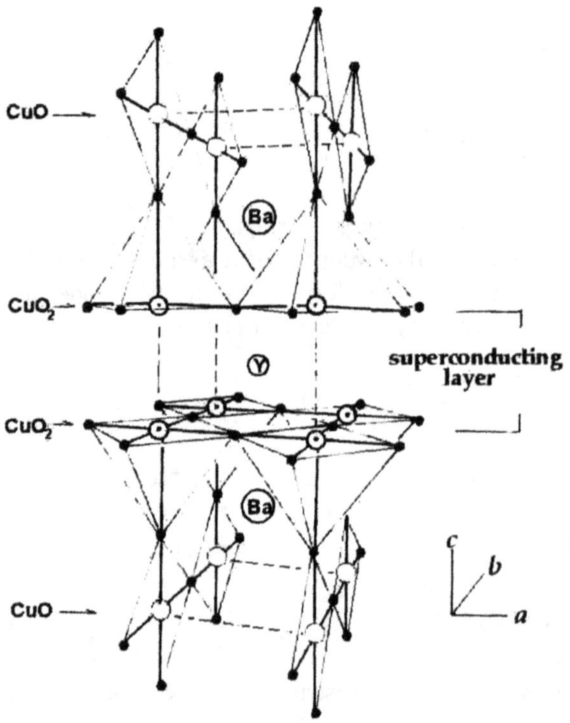

Figure 11.6. The cuprate-layer structure with Y ions in-between of $YBa_2Cu_3O_{7-x}$ lattice. This image has been obtained by the author from http://en.wikipedia.org/wiki/High-temperature_superconductivity, where it is stated to have been released into the public domain. It is included within this chapter on this basis.

Table 11.1. Critical temperatures of cuprate layer structures[a]

Formula (Notation)	T_c (K)	Number of cuprate planes in unit cell
$YBa_2Cu_3O_7$ (123)	92	2
$Bi_2Sr_2CuO_6$ (Bi2201)	20	1
$Bi_2Sr_2CaCu_2O_8$ (Bi–2212)	85	2
$Bi_2Sr_2Ca_2Cu_3O_6$ (Bi–2223)	110	3
$Tl_2Ba_2CuO_6$ (Tl–2201)	80	1
$Tl_2Ba_2CaCu_2O_8$ (T1–2212)	108	2
$T1_2Ba_2Ca_2Cu_3O_{10}$ (Tl–2223)	125	3
$HgBa_2CuO_4$ (Hg–1201)	94	1
$HgBa_2CaCu_2O_6$ (Hg–1212)	128	2
$HgBa_2Ca_2Cu_3O_8$ (Hg–1223)	134	3

[a] Data from http://en.wikipwdia.org//wild//High-temperature_superconductivity.

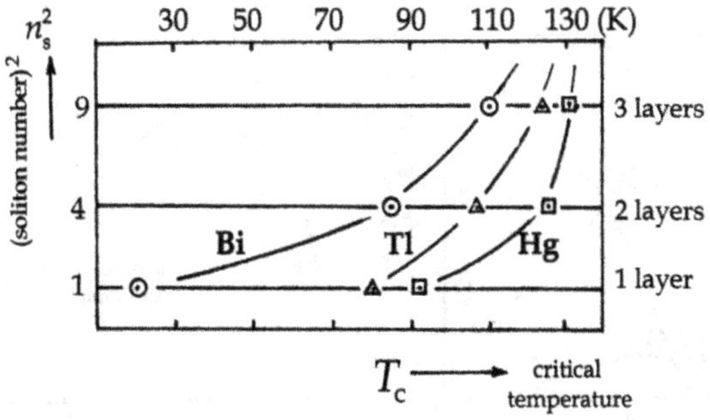

Figure 11.7. Verifying the formula $T_c \propto n_s^2$ from the soliton theory with data published in reference [3].

Looking at table 11.1 published in Wikipedia, there is data on the observed transition temperatures from cuprate-layer compounds, which could be analyzed by the solution theory. Figure 11.7 shows the results, which provide clear evidence to support equation (11.3*h*) to be determined by $T_c \propto n_s^2$, i.e. proportional to the square of solution number n_s, while uncertain in experimental accuracy of reference [3]. Nonetheless, that was no obvious result from the BCS theory.

11.7 Meissner's effect in cuprate-layers and metallic hydrogen sulfide H3S

Figure 11.8(*b*) illustrates experimental results by Eremets and his group [2], showing that the transition in H3S under high external pressures was not as sharp as in isothermal cases. Similar results were also reported on magnetic studies by Tarnawski [5] on high-T_c superconductors $RBa_2Cu_3O_{7-x}$, as shown in figure 11.9(*a*), indicating a similar transition to Type 2 superconductors [6]. The physical reason for these

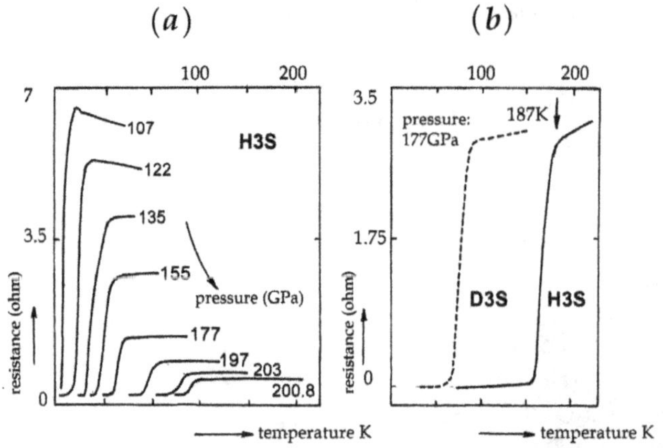

Figure 11.8. (*a*) Zero-resistance curves observed from H3S at various values of applied pressure. (*b*) A comparison of zero-resistance experiments between H3S and D3S at 177 GPs and 185 K. Data from [2].

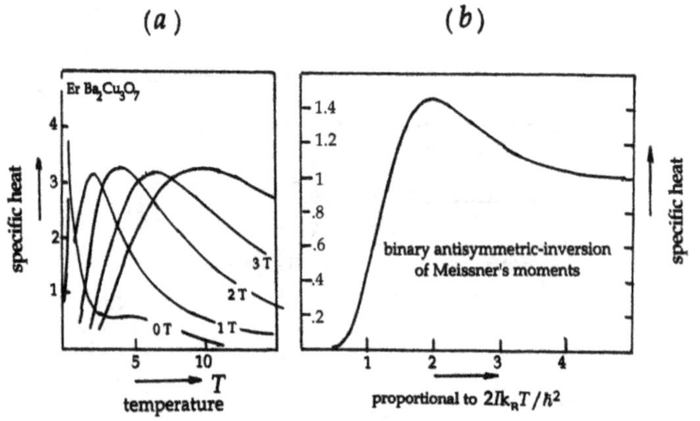

Figure 11.9. (*a*) Transition anomalies in superconducting $ErBa_2Cu_3O_7$ for $T > T_c$ [5]. (*b*) Specific heat anomaly calculated for antisymmetric Meissner's pairs as a function of applied $\boldsymbol{H}$ [6].

magnetic observations was obviously attributed to soliton correlations with the *elastic crystalline environments*, while appropriate analyses have not yet been carried out. Nevertheless, it is interesting to realize that the curves of specific heat from a gaseous material of Meissner's *magnetic dipolar molecules* in applied magnetic fields are found to be similar to *para-hydrogen gas*, referring to Wannier's textbook of *Statistical Physics* [7]. Basically consistent with Δz_P and Δz_A in (11.3*c*) as indicated in figure 11.9(*b*) for the initial transition, the phase above the transition point is determined by *anti-symmetrical lattice displacements*, like in the case of gaseous *para-hydrogen.*

Figure 11.8(*a*) shows some of the observations published for *zero-resistance of protonic currents* by Eremets and his associates [2]. Indicating in figure 11.3 that protons 1 are responsible for normal currents, figure 11.8(*a*) shows evidence that the normal current $\boldsymbol{j}_n$ is converting to persistent diamagnetization $\boldsymbol{M} = \mathrm{curl}\,\boldsymbol{j}_s$, which

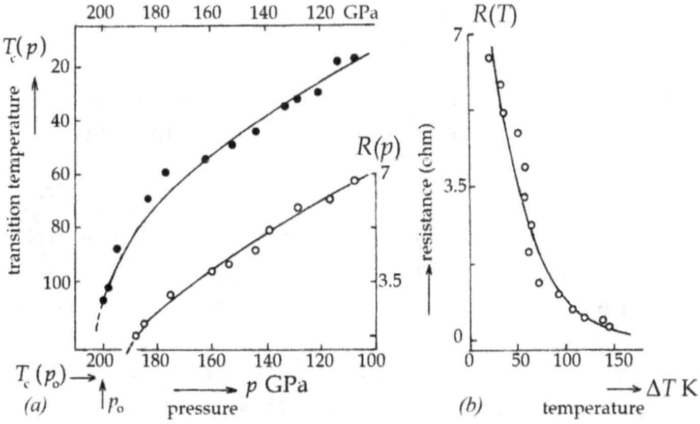

Figure 11.10. Examples of soft-mode behavior. (*a*) $T_c(p)$ plotted as a function of applied pressure p, showing proportional to soliton density $n_s(p)$ that is proportional to $\sqrt{p_0 - p}$ in the mean-field approximation. (*b*) The normal resistances $R(T) \propto (T - T_c)^{-1}$ and $R(p) \propto (p_0 - p)^{-1}$ exhibited similar Curie–Weiss anomalies, as shown.

was not explicit in the resistant measurements, but obeying the charge-current continuity law over the *adiabatic transition region*. Those experimental results in figure 11.9 in the presence of external pressure are only sufficiently accurate, as analyzed as illustrated in figure 11.9, where such a *parabolic curve* as $T_c(p) \propto \sqrt{p_0 - p}$ exhibits soft mode and Curie–Weiss' behavior of $R(T)$ above T_c near the critical point $T_c = T_{p_0}$, supporting the soliton theory.

Although unformulated, resistance measurements constitute thermodynamic processes, hence we can analyze the measuring process with respect to Anderson's theory. Considering (11.6*e*), we write

$$\langle \mathcal{H}_{\mathrm{BCS}} \rangle_{\mathrm{thermal}} = -\langle \varepsilon_k \rangle \sigma_{kz} - \sigma_{kz} F_{kz}$$

where F_{kz} represents internal Weiss' field, and $\langle \varepsilon_k \rangle$ is the average protonic kinetic energy. And the external pressure energy $-p\, A_\perp \Delta z$ is assumed for the response from the protonic energy $\Delta \langle \varepsilon_k \rangle = \frac{\partial \langle \varepsilon_k \rangle}{\partial z} \Delta z$. Accordingly, we have a thermodynamic relation in the presence of external high pressure, i.e.

$$\Delta G(\sigma_z; \Delta p) = -\frac{\partial \langle \varepsilon_k \rangle}{\partial z} (\Delta z) \sigma_z + \sigma_z A_\perp \Delta p. \tag{11.7}$$

Setting $\Delta G(\sigma_z, \Delta p) = 0$ for minimum, we have the relation for the modulated $(\Delta z)\sigma_z$ to be studied with varying external pressure Δp, as illustrated by figure 11.10.

All in all, the soliton theory of superconductivity is well-established on these crystalline conducting systems, as substantiated experimentally.

Exercises

1. For superconductivity, the Fröhlich interaction is for binary inversion of the soliton. Therefore, it should be for an adiabatic interaction with an inductive

potential $\pm i\frac{du}{dx}$. Confirm that idea from our discussion of critical potential energy in this chapter.

2. The Cooper pair is basically equivalent to cluster formation in conventional binary transition. Confirm that from discussions in this chapter.

References

[1] Kittel C 1986 *Introduction to Solid State Physics* 6th ed (New York: Wiley)
Fujimoto M 2017 *Solitons in Crystalline Processes* (Bristol: IOP Publishing)

[2] Drozdov A P, Eremets M I and Troyan I A 2014 Conventional superconductivity at 190 K at high pressure arXiv: 1412.0460

[3] Wikipedia http://en.wikipedia.org/wiki/High-temperature_superconductivity

[4] Kittel C 1963 *Quantum Theory of Solids* (New York: Wiley) ch 8

[5] Tarnawski Z 1996 *Mol. Phys. Rep.* **15/16** 103

[6] Kittel C 1956 *Introduction to Solid State Physics* 3rd ed (New York: Wiley) ch 11

[7] Wannier G H 1966 *Statistical Physics* (New York: Wiley)

[8] Einaga M, Sakata M, Ishikawa T, Shimizu K, Eremets M I, Drozdov A P, Troyan I A, Hirano N and Ohishi Y 2016 *Nat. Phys.* **12** 835

IOP Publishing

Introduction to the Mathematical Physics of Nonlinear Waves
(Second Edition)

Minoru Fujimoto

Chapter 12

Irreducible thermodynamics of superconducting phase transitions

Superconducting transitions take place in crystals that are signified by conduction bands of free electrons or protons, interacting with the structural change at the critical point of phase transitions in a specific type. Accordingly, zero-resistance phenomena were discovered as characterized by Meissner's effect that was explained by Fröhlich's lattice distortion. On the other hand, a hypothetical Cooper pair was replaced by the soliton theory, all the existing consequences remain unchanged with some corrections. In this chapter, we review existing theories of the superconducting phase with respect to traditional irreversible thermodynamics.

12.1 Superconducting phase transition

12.1.1 Meissner's diamagnetism and the persistent current

Historically, the superconductivity recognized by *zero-resistance experiments* was observed for a small magnet to float in the air up-above superconducting surfaces below transition temperatures. Meissner discovered that such phenomena occurred with an applied magnetic field $\boldsymbol{H}$. In this case, we should consider that the magnetic image force is responsible for lifting the magnet.

Referring to figure 11.1 in chapter 11 that illustrates a nonmagnetic sample crystal in the applied magnetic field $\boldsymbol{H}$, where figure 11.1(*a*) shows that nothing occurs at temperatures $T > T_c$, whereas figure 11.1(*b*) for the superconducting phase $T < T_c$, the flux of magnetic induction $\boldsymbol{B} = \mu_o \boldsymbol{H}$ are all expelled out of the sample, thereby characterizing the phase by $\boldsymbol{B} = 0$ or $\boldsymbol{M} = \mu_o \boldsymbol{H}$ where $\mu_o = 4\pi \times 10^{-7}$ V (A m)$^{-1}$ is a constant of vacuum space. Figure 11.1(*c*) shows the flux pattern generated from the superconducting sample at $\boldsymbol{B} = 0$.

doi:10.1088/978-0-7503-3759-5ch12

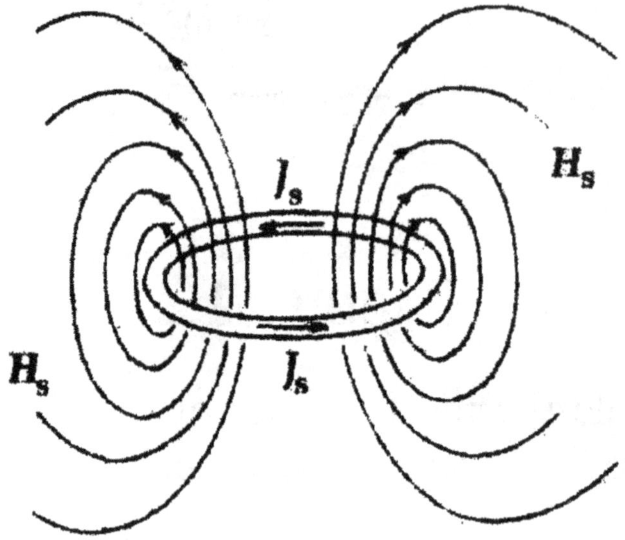

Figure 12.1. A superconducting ring for $T < T_c$ after removing applied $\boldsymbol{B}$. The distributed flux represents a trapped field due to the persistent current curl $\boldsymbol{j}_s = \boldsymbol{H}$.

Figure 12.1 sketches Collins' experiment [1] on a superconducting magnet, substantiating the *law of charge–current continuity in Maxwell's electromagnetism* and Biot–Savart's theorem, thereby verifying the relation between supercurrent $\boldsymbol{J}_s$ and Meissner's diamagnetism $\boldsymbol{M}$, exhibiting $\boldsymbol{M}$ = curl $\boldsymbol{J}_s$. Therefore, we can define a magnetic energy contribution $\boldsymbol{M} \cdot \boldsymbol{H}$ per volume to the Gibbs function of a superconductor. Accordingly, thermodynamically, the applied magnetic field $\boldsymbol{H}$ is an external variable like applied pressure p.

12.1.2 Thermodynamics of a superconducting transition

Thermodynamically, the order variable is the Cooper pair of *two fermions*, which are free moving particles. Therefore, the internal energy can basically be determined by Sommerfeld's model whose specific heat is proportional to T^3 at low temperatures, accompanied by the phonons' contribution from the lattice. Accordingly, the phase transition is signified by its specific-heat *anomaly* $\Delta C_{\text{Cooper-pair}}$ characterized as λ-transition, showing empirically for

$$\Delta C_{\text{Cooper pair}} \sim e^{-b/T}, \tag{12.1}$$

where the constant is related to an energy gap between normal and superconducting states at T_c.

As indicated by Meissner's effect, the superconducting state is signified by diamagnetization $\boldsymbol{M} = -\mu_0 \boldsymbol{H}$ by external magnetic field $\boldsymbol{H}$ for $T < T_c$, while the normal state is specified by $\boldsymbol{B} \neq 0$. On the other hand, the transition point can be specified by T_c and H_c. Figures 12.2(a) and (b) show experimental results obtained

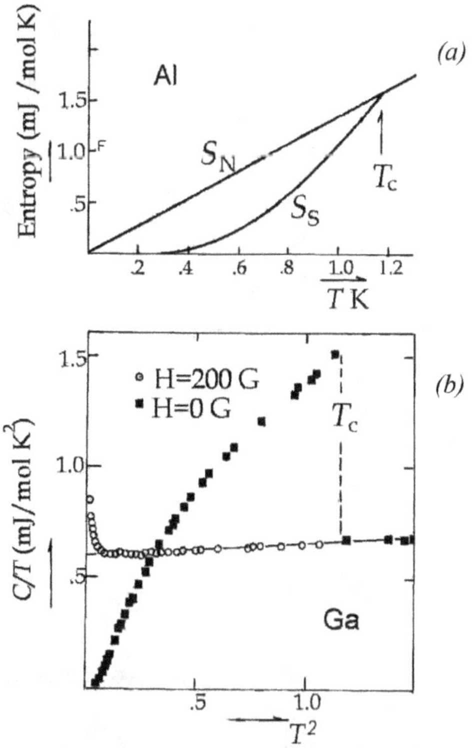

Figure 12.2. (*a*) Entropy S vs temperature curves for coexisting normal- and superconducting phases of metallic Al for $T < T_c$. (*b*) C_V/T vs T^2 curves for Ga. Superconducting phase transition occurs when $H = 0$, is compared with no transition when $H = 200$ G. Reproduced with permission from [2]. © 1986 John Wiley & Sons.

from Ga metals [2], demonstrating detail of the phase transitions composed of normal and superconducting phases below T_c, showing that their data from Ga,

$$C_V/T = 0.596 + 0.0586\, T^{-2},$$

was close to the theoretical assumption for $\Delta C_{\text{Cooper pair}}$.

For a bulk of superconducting phase in ellipsoidal shape, Meissner considered that the body has the magnetic energy density $\boldsymbol{M} \cdot \boldsymbol{H}$ per unit-volume in a given $\boldsymbol{H}$. Therefore, the work to increase the magnetic energy can be expressed by $\boldsymbol{H} \cdot \mathrm{d}\boldsymbol{M}$ per volume, and the first law of thermodynamics is formulated as

$$\mathrm{d}U = T\mathrm{d}S - p\mathrm{d}V + \boldsymbol{H} \cdot \mathrm{d}\boldsymbol{M}. \tag{12.2}$$

Defining Gibbs' free energy $G = U + \mathrm{d}U$, we can set up the equilibrium condition for minimizing $\mathrm{d}G \geqslant 0$. Then, the equilibrium between normal and superconducting phases can be written as

$$G_s(H_c) = G_n(H_c), \quad G_s(0) = G_n(0) \tag{12.3a}$$

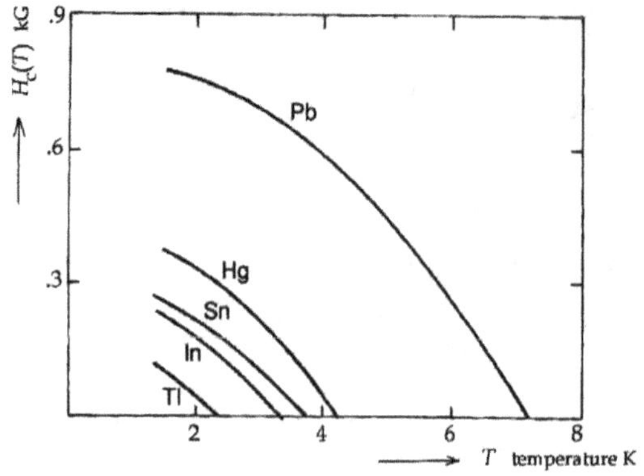

Figure 12.3. H_c vs T diagrams, comparing several superconducting metals. Reproduced with permission from [2]. © 1986 John Wiley & Sons.

and

$$G_s(H_c) = G_n(0) - \mu_o \int_0^{H_c} H \, dH = G_n(0) - \frac{\mu_o H_c^2}{2}. \tag{12.3b}$$

Therefore, the difference between entropies of these phases is derived from (12.3*b*) as

$$S_n - S_s = -\frac{d(G_n - G_s)}{dT} = -\mu_o H_c \frac{dH_c}{dT}, \tag{12.3c}$$

and the related latent heat at the transition temperature is expressed as

$$L = T(S_n - S_s) = -TH_c \frac{dH_c}{dT}. \tag{12.3d}$$

Figure 12.3 shows the plots obtained from anomalies in metallic superconductors for H_c against temperatures T, characterized by their *parabolic* curves. The specific heat measurement as expressed by

$$\Delta C_V = \mu_o T \frac{\partial(S_n - S_s)}{\partial T} = \mu_o T \left\{ H_c \frac{d^2 H_c}{dT^2} + \left(\frac{dH_c}{dT} \right)^2 \right\},$$

indicating a specific form

$$(\Delta C_V)_{T_c} \approx \mu_o T_c \left(\frac{dH_c}{dT} \right)_{T_c}^2 \to 0, \text{ because of } \lim_{T_c \to 0} \frac{dH_c}{dT_c} \to 0,$$

which represents the third law of thermodynamics, while we cannot confirm experimentally, otherwise confirmed by the appearance of a supercurrent $\boldsymbol{J}_s$ for curl $\boldsymbol{J}_s \neq 0$.

The anomalies shown by figure 12.3 clearly exhibit soft-mode behavior originating from soliton dynamics, as analyzed to be related with the Meissner effect (12.2).

In the temperature range closer to $H_c(T_c) \to 0$, the curves are almost linear to ΔT_c, hence

$$H_c(\Delta T_c) \propto \sqrt{\Delta T_c} \tag{12.3e}$$

can be consistent with the *soliton anomaly*, known as *soft-mode behavior*. In fact, those critical values M_c, H_c and T_c are related with the law of ideal soliton gas

$$M_c H_c \propto T_c \tag{12.3f}$$

at the critical point when adiabatic processes are disregarded.

12.2 Electromagnetic properties of superconductors

12.2.1 Persistent current

In early physics, zero-resistance was idealized by infinite electric conductivity, but the superconductivity was revised by perfect diamagnetism after the discovery of persistent current. In any case, the latter was found to occur as related to soliton mobility in the momentum space. Accordingly, the law of charge–current continuity was applied to Cooper pairs, as consistent with the Maxwell theory of electromagnetism.

Hence, we can characterize the supercurrent to flow by

$$\Lambda \frac{\partial \boldsymbol{J}_s}{\partial t} = \boldsymbol{E}_s, \tag{12.4a}$$

where Λ is a characteristic constant and $\boldsymbol{E}_s = -\nabla_r V(\boldsymbol{r}, t)$ where $V(\boldsymbol{r}, t)$ is Fröhlich's potential, existing for $T < T_c$. In contrast, the normal current of single carrier obeys the Ohm law, being proportional to applied electric field $\boldsymbol{E}_{ext}$, as expressed by

$$\boldsymbol{J}_n = \sigma \boldsymbol{E}_{ext}, \tag{12.4b}$$

where σ is conductivity of a crystal for $T > T_c$.

Writing the charge and mass of the Cooper pair as $e' = 2e$ and m', respectively, $\boldsymbol{J}_s = n_s e' \boldsymbol{v}_s$ with the *soliton number* n_s and the equation of motion $m' \dot{\boldsymbol{v}}_s = e' \boldsymbol{E}_s$. Then, from (12.3*a*) we can derive the relation

$$\Lambda = \frac{m'}{n_s e'^2}. \tag{12.4c}$$

12.2.2 Penetration depth

Using (12.4*a*) for supercurrent, we can shows that the magnetic induction $\boldsymbol{B}$ exists only in the surface area in finite depth, leaving inside as $\boldsymbol{B} = 0$.

First, we can accommodate the supercurrent $\boldsymbol{J}_s$ and related $\boldsymbol{E}_s$ in basic Maxwell's equations, i.e. we have

$$\text{curl}\, \boldsymbol{E}_s = -\frac{\partial \boldsymbol{B}}{\partial t} \quad \text{and} \quad \text{curl}\, \boldsymbol{B} = \mu_0 \left(\boldsymbol{J}_s + \varepsilon_0 \frac{\partial \boldsymbol{E}_s}{\partial t} \right).$$

Combining the first equation with (12.4*a*), we obtain curl $\left(\Lambda\frac{\partial \boldsymbol{J}_s}{\partial t}\right) = -\frac{\partial \boldsymbol{B}}{\partial t}$. Assuming $\frac{\partial E_S}{\partial t} = 0$ for a steady current $\boldsymbol{J}_s$, the second Maxwell's equation can be reduced to curl $\boldsymbol{B} = \mu_o \boldsymbol{J}_s$, indicating that $\boldsymbol{B}$ is distributed with supercurrent $\boldsymbol{J}_s$, as calculated from

$$\frac{\partial \boldsymbol{B}}{\partial t} = -\text{curl}\left(\frac{\Lambda}{\mu_o}\text{curl}\frac{\partial \boldsymbol{B}}{\partial t}\right) = -\lambda^2 \,\text{curl}\,\text{curl}\frac{\partial \boldsymbol{B}}{\partial t} = -\lambda^2\nabla^2\frac{\partial \boldsymbol{B}}{\partial t} \quad \text{where} \quad \lambda^2 = \frac{\Lambda}{\mu_o}.$$

Hence, especially for the field $\boldsymbol{B} = (0, 0, B_z)$ to interpret $\lambda = \lambda_d$, we have the relation

$$\lambda_d{}^2\nabla^2\dot{B}_z - \dot{B}_z = 0, \tag{12.5a}$$

whose solution is given by $\dot{B}_z = \dot{B}_o e^{-z/\lambda_d}$ with a *penetration depth* λ_d.

It is important to realize at this point that the supercurrent plays an essential role for the penetration to verify the Meissner effect, which cannot be explained only by Maxwell's theory. Hence, London revised (12.5*a*) as

$$\boldsymbol{B} = \lambda_d{}^2\nabla^2\boldsymbol{B} \quad \text{with his formula curl}(\Lambda\boldsymbol{J}_s) = -\boldsymbol{B}. \tag{12.5b}$$

Based on (12.5*b*), Meissner's effect can be confined to surface areas, but the theory is qualitative in principle, because fabricated sample crystals are not usually ellipsoidal, and the depth $\lambda_d \sim 10^{-5}$ cm gives only the order of magnitude.

12.2.3 London's gauge function and magnetic flux quantization

In the absence of magnetic charges, the magnetic induction vector $\boldsymbol{B}$ can be expressed by the relation div $\boldsymbol{B} = 0$, hence determined by its related vector potential $\boldsymbol{A}$ as

$$\boldsymbol{B} = \text{curl}\,\boldsymbol{A}. \tag{12.6a}$$

However, the vector potential $\boldsymbol{A}$ cannot be uniquely determined, accompanying by arbitrary function χ as

$$\boldsymbol{A} \to \boldsymbol{A} + \nabla\chi \quad \text{where} \quad \nabla^2\chi = 0. \tag{12.6b}$$

Here, equation (12.6*b*) is called a *gauge transformation*, where χ is an arbitrary scalar function.

Using $\boldsymbol{A}$, London's equation (12.5*b*) is expressed as

$$\text{curl}(\Lambda\boldsymbol{J}_s + \boldsymbol{A}) = 0,$$

therefore

$$\Lambda\boldsymbol{J}_s + \boldsymbol{A} = \nabla\chi' \quad \text{where} \quad \nabla^2\chi' = 0. \tag{12.6c}$$

Here, χ' is another arbitrary function, but like χ in (12.6*b*), so we consider they are the same arbitrary function, but in the presence of $\boldsymbol{J}_s$, χ' is called *London's gauge* in particular. For (12.6*c*), we have

$$\boldsymbol{A} = -\Lambda\boldsymbol{J}_s, \tag{12.6d}$$

representing the participation of the supercurrent together with motion of Cooper pairs in the momentum space.

For a system of independent superconducting Cooper pairs, we can write a Hamiltonian

$$\mathcal{H}_{\text{system}} = \sum_i \frac{\{\boldsymbol{p}_s(\boldsymbol{r}_i) - e'\boldsymbol{A}(\boldsymbol{r}_i)\}^2}{2m'} = \int_{\text{volume}} \mathcal{H}(\boldsymbol{r})\mathrm{d}^3\boldsymbol{r},$$

where

$$\mathcal{H}(\boldsymbol{r}) = \frac{\{\boldsymbol{p}_s(\boldsymbol{r}) - e'\boldsymbol{A}(\boldsymbol{r})\}^2}{2m'} \tag{12.7a}$$

is a one-particle Hamiltonian density. And the vector potential in invariant under a gauge transformation

$$\boldsymbol{A}(\boldsymbol{r}') = \boldsymbol{A}(\boldsymbol{r}) + \nabla\chi(\boldsymbol{r}) \quad \text{where} \quad \nabla^2\chi(\boldsymbol{r}) = 0. \tag{12.7b}$$

Hence, for the invariance of between eigen-equations

$$\mathcal{H}(\boldsymbol{r})\psi(\boldsymbol{r}) = E\psi(\boldsymbol{r}) \quad \text{and} \quad \mathcal{H}(\boldsymbol{r}')\psi(\boldsymbol{r}') = E\psi(\boldsymbol{r}'),$$

we need to consider that

$$\psi(\boldsymbol{r}') = e^{\frac{ie'\chi}{\hbar}}\psi(\boldsymbol{r}), \tag{12.7c}$$

indicating only a phase shift $e'\chi/\hbar$.

Accordingly, considering the presence of n_s Cooper pairs, the total momentum of the particles should be expressed as

$$\boldsymbol{P}_s(\boldsymbol{r}) = n_s(\boldsymbol{p}_s(\boldsymbol{r}) - e'\boldsymbol{A}(\boldsymbol{r})),$$

satisfying the relation

$$\text{curl}\, \boldsymbol{P}_s(\boldsymbol{r}) = 0, \quad \text{and hence} \quad \boldsymbol{P}_s(\boldsymbol{r}) = n_s\nabla\chi(\boldsymbol{r}) \tag{12.7d}$$

where the gauge function $\chi(\boldsymbol{r})$ is a scalar, and

$$\boldsymbol{P}_s(\boldsymbol{r}) = 0 \tag{12.7e}$$

in a simply connected conductor, whereas in a *multi-connected conductor*, special consideration needs to deal with distributed $\boldsymbol{P}_s(\boldsymbol{r})$.

In a multi-connected conductor however, $\chi(\boldsymbol{r})$ cannot take a unique value at a given position. Figure 12.4 sketches a superconducting domain surrounded by a normal phase that is shaded in the figure. In a practical metal, the body may have a void bordered by Σ, which can be either in normal phase or an empty space. In the figure, inside the closed curve S_1, the metal is superconducting, whereas S_2 contains multiply connected space.

Since (12.7*d*) is written as a function of local point $\boldsymbol{r}$, the surface integral on S_1 shows $\int_{S_{1+}} \boldsymbol{P}_s \cdot \mathrm{d}\boldsymbol{S}_{1+} = \int_{S_{1-}} \boldsymbol{P}_s \cdot \mathrm{d}\boldsymbol{S}_{1-}$ where $\mathrm{d}\boldsymbol{S}_{1+}$ and $\mathrm{d}\boldsymbol{S}_{1-}$ represent the surface

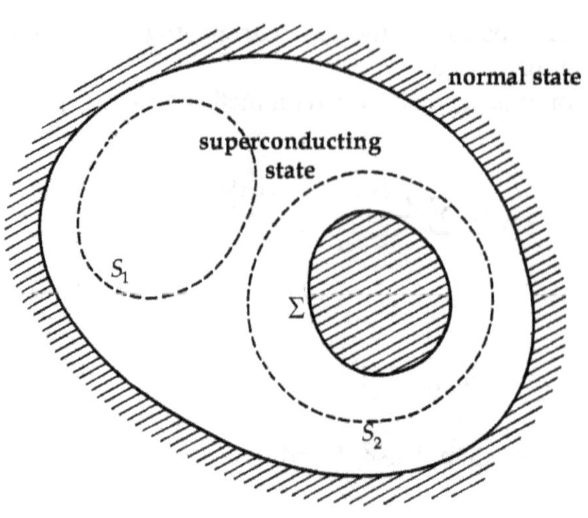

Figure 12.4. Superconducting state in multi-connected phase.

elements inside and outside, respectively, verifying London's equation (12.6*c*) over the space S_1.

A supercurrent coil as sketched in figure 12.1 can be analyzed with the space S_2 in figure 12.4. Persisting currents in a superconducting coil last a considerably long time. As evidenced by such persisting currents, Meissner's effect can be re-phrased for a multi-connected body, regarding to trapped flux of magnetic induction threaded through the border hole Σ.

For the space including Σ to consider as a doubly connected body, two integrals $\oint_{S_2} \boldsymbol{p}_s \cdot \mathrm{d}\boldsymbol{S}_2$ and $\oint_\Sigma \boldsymbol{p}_s \cdot \mathrm{d}\boldsymbol{\Sigma}$ can be calculated, but the former vanishes, and the latter determines the total integral. In this case, it is convenient to use the magnetic induction $\boldsymbol{B}$ to calculate the flux trapped inside of Σ.

Hence, we use the formula for flux of $\boldsymbol{B}$

$$-\frac{\partial}{\partial t}\oint_\Sigma \boldsymbol{B} \cdot \mathrm{d}\boldsymbol{\Sigma} = \oint_\Sigma \mathrm{curl}\, \boldsymbol{E}_s \cdot \mathrm{d}\boldsymbol{\Sigma} = \oint_\Sigma \boldsymbol{E}_s \cdot \mathrm{d}\boldsymbol{l}_\Sigma, \tag{12.8a}$$

where the surface penetration of $\boldsymbol{B}$ on Σ is ignored for simplicity, and Stokes theorem is applied to obtain the last line integral. Accordingly, equation (12.8) can be written as

$$\frac{\partial}{\partial t}\left(\oint_\Sigma \boldsymbol{B} \cdot \mathrm{d}\boldsymbol{\Sigma} + \Lambda \oint_\Sigma \boldsymbol{J}_s \cdot \mathrm{d}\boldsymbol{l}_\Sigma\right) = 0, \tag{12.8b}$$

allowing to define the total flux as

$$\Phi = \oint_\Sigma \boldsymbol{B} \cdot \mathrm{d}\boldsymbol{\Sigma} + \Lambda \oint_\Sigma \boldsymbol{J}_s \cdot \mathrm{d}\boldsymbol{l}_\Sigma = \oint_\Sigma (\boldsymbol{A} + \Lambda \boldsymbol{J}_s) \cdot \mathrm{d}\boldsymbol{l}_\Sigma. \tag{12.8c}$$

Therefore (12.8*b*) indicates $\Phi = \text{const.}$, whose value is determined from (12.8*c*) by the gauge $\chi(\boldsymbol{r})$, i.e.

$$\Phi = \oint_{\Sigma} \nabla\chi(\boldsymbol{r}) \cdot \mathrm{d}\boldsymbol{l}_{\Sigma} = \frac{2\pi\hbar\, n_{\mathrm{s}}}{e'} = n_{\mathrm{s}}\chi' \tag{12.8d}$$

where $\chi' = \frac{2\pi\hbar}{e'}$ is the London gauge for the space coordinate $\boldsymbol{r}$ to consider as a thermodynamic variable. Equation (12.8*d*) indicates that the *fluxoid* Φ is quantized as proportional to the soliton number n_s, showing the electromagnetic properties are compatible with the soliton concept.

12.3 The Ginzburg–Landau equation for superconducting phase transitions

For thermodynamic description of Cooper pairs in superconductors, Ginzburg and Landau [3] wrote Gibbs' free energy density as follows.

$$g_{\mathrm{s}} = g_{\mathrm{n}} + \frac{1}{2m'}\{(\boldsymbol{p}_{\mathrm{s}} - e'\boldsymbol{A})\psi\}^2 + \alpha\,|\psi|^2 + \frac{1}{2}\beta(|\psi|^2)^2 - \frac{1}{2\mu_{\mathrm{o}}}\boldsymbol{M}^2, \tag{12.9a}$$

where the soliton density is proportional to the pair's density, and given by

$$n_{\mathrm{s}} = |\psi|^2, \tag{12.9b}$$

corresponding to the soliton number to form the Weiss potential. Here, variations $\delta\psi$ and $\delta\boldsymbol{A}$ are considered to represent Cooper pairs and supercurrents, respectively for the superconducting transition to exhibit normal and superconducting phases specified by suffixes *n* and *s*.

Therefore, by minimizing $g_{\mathrm{s}} - g_{\mathrm{n}}$ with respect to $\delta\psi$ and $\delta\boldsymbol{A}$, namely

$$\delta(g_{\mathrm{s}} - g_{\mathrm{n}}) = \left(\frac{\partial g_{\mathrm{s}}}{\partial\psi}\right)_{\boldsymbol{A}}\delta\psi + \left(\frac{\partial g_{\mathrm{s}}}{\partial\boldsymbol{A}}\right)_{\psi}\delta\boldsymbol{A} = 0$$

for $\left(\frac{\partial g_{\mathrm{s}}}{\partial\psi}\right)_{\boldsymbol{A}} = 0$ and $\left(\frac{\partial g_{\mathrm{s}}}{\partial\boldsymbol{A}}\right)_{\psi} = 0$ to determine equilibrium conditions. Therefore, we obtain the relations

$$\frac{1}{2m'}(-i\hbar\nabla - e'\boldsymbol{A})^2\psi + (\alpha + \beta\,|\psi|^2)\psi = 0 \tag{12.10a}$$

and

$$\boldsymbol{J}_{\mathrm{s}}(\boldsymbol{r}) = -\frac{i\hbar\, e'}{2m'}(\psi^*\nabla\psi - \psi\nabla\psi^*) - \frac{e'^2}{m'}\psi^*\psi\boldsymbol{A}, \tag{12.10b}$$

respectively. Equation (12.10*a*) is known as the *Ginzburg–Landau equation*, and (12.10*b*) provides the supercurrent density.

Equation (12.10*a*) can be solved in perturbational manner, if considering the transition is initiated by externally applied vector potential $\boldsymbol{A}$ that is proportional to supercurrent. In this case, the unperturbed wave equation is initiated by the linear potential $\alpha\psi$, becoming then nonlinear by the $\beta\,|\psi|^2\psi$ as determined by

$$\frac{\hbar^2}{2m'}\frac{d^2\psi}{dx^2} + \alpha\psi + \beta\,|\psi|^2\psi = 0 \tag{12.10c}$$

where $\alpha < 0$ and $\beta > 0$ for $T < T_c$. Thereby, the initial harmonic wave function $e^{ix/\xi}$ close to T_c, where $\xi = \hbar/\sqrt{2m'|\alpha|}$, becomes a nonlinear function $\psi(x) = \sqrt{\frac{|\alpha|}{\beta}}$ $\tanh\frac{x}{\sqrt{2}\xi}$. Considering $|\psi|^2 = \frac{|\alpha|}{\beta}$, we obtain $g_s - g_n = -\frac{\alpha^2}{2\beta} = -\frac{1}{2}\mu_o H_c^2$, representing thermodynamically an ordered state of the superconductor. Therefore, we have the expression for the critical magnetic field for superconductor of Type 1

$$H_c = \frac{|\alpha|}{\sqrt{\mu_o\beta}}. \tag{12.10d}$$

In contrast, equation (12.10*b*) indicates that the supercurrent exists in the presence of applied $\boldsymbol{A}$, even if $\nabla\psi$ is negligibly small, the current density

$$\boldsymbol{J}_s(\boldsymbol{r}) = -\frac{e'^2}{m'}\,|\psi|^2\boldsymbol{A}(\boldsymbol{r}) \tag{12.10e}$$

is present in the penetration depth determined by $\lambda_d = \sqrt{\frac{m'\beta}{\mu_o e'^2|\alpha|}}$. In addition, equation (12.10*b*) applied for the transition threshold, we have

$$\frac{1}{2m'}(-i\hbar\nabla - e'\boldsymbol{A})^2\psi + \alpha\psi = 0. \tag{12.10f}$$

If the field $\boldsymbol{B}$ is applied parallel to the y-direction, equation (12.10*f*) is written as

$$-\frac{\hbar^2}{2m'}\left(\frac{\partial^2}{\partial x^2} + \frac{\partial^2}{\partial z^2}\right)\psi + \frac{1}{2m'}\left(i\hbar\frac{\partial}{\partial y} + e'B\right)^2\psi = \alpha\psi,$$

which can be modified for $\psi = \psi_o e^{ik_z z}$ as

$$\frac{1}{2m'}\left(-\hbar^2\frac{d^2}{dx^2} + \hbar^2k_z^2 + (\hbar\,k_z - e'Bx)^2\right)\psi_o = \alpha\psi_o,$$

and then

$$\frac{1}{2m'}\left(-\hbar^2\frac{d^2}{dx^2} + e'^2B^2x^2 - 2\hbar\,k_z e'Bx\right)\psi_o = \left(\alpha - \frac{\hbar^2\left(k_y^2 + k_z^2\right)}{2m'}\right)\psi_o.$$

Applying a coordinate transformation $x - \frac{\hbar\,k_y e'B}{2m'} \to x'$, we obtain

$$\left(-\frac{\hbar^2}{2m'}\frac{d^2}{dx'^2} - \frac{1}{2}m'\omega_L^2 x'^2\right)\psi_o = \left(\alpha - \frac{\hbar^2k_z^2}{2m'}\right)\psi_o, \tag{12.10g}$$

where $\omega_L = \frac{e'B}{m'}$ is the Larmor frequency at $\boldsymbol{B}$, and the lowest eigenvalue is $\alpha = \frac{1}{2}\hbar\omega_L$, if $k_z = 0$. Therefore (12.10g), can be studied by magnetic spectroscopy, however thermal measurements of specific anomaly can determine critical temperature for simplicity. In most cases, like superconductivity of type 2 where nonlinearity is intensive, and critical anomalies do not occur sharply [4]. As related by many other factors than α, the critical magnetic field of type 2 can be estimated by

$$H_{c2} = \frac{\sqrt{2}\,\lambda}{\xi}H_{c1}. \tag{12.10h}$$

At this stage, it is noted that in multi-connected superconductors, such magnetic field H_{c2} is quantized as shown below, since the magnetic flux $\Phi = \oint_\Sigma \boldsymbol{B}_{c2}\cdot d\boldsymbol{\Sigma} = n_s\Phi_0$ is trapped inside voids Σ. Namely,

$$H_{c2} = \frac{2m'\alpha}{e'\hbar}\frac{e'\Phi_0}{2\pi\hbar}\frac{\hbar^2}{2m'\xi^2} \propto \frac{n_s\,\Phi_o}{2\pi\xi^2}. \tag{12.10i}$$

12.4 Field theory of superconducting transitions

12.4.1 Bardeen–Cooper–Schrieffer's ground states

The field theory of superconducting states of multi-electron or proton systems is supported for Cooper pairs by the soliton theory, so that the existing Bardeen–Cooper–Schrieffer (BCS) theory does not need any significant revisions. Nevertheless, these theories are statistical and compatible with thermodynamic principles, as summarized in this section.

Writing the BCS Hamiltonian as

$$\mathcal{H}_{BCS} = \sum_k \mathcal{H}_k = \sum_k \varepsilon_k(a_k^\dagger a_k + a_{-k}^\dagger a_{-k}) - V\sum_k a_k^\dagger a_{-k}^\dagger a_{-k} a_k \tag{12.11a}$$

where these $a_k^\dagger$ and a_k are quantum mechanical creation- and anhelation-operators of Fermions, and their time-variations are determined by Heisenberg's relations

$$i\hbar\frac{\partial a_k^\dagger}{\partial t} = [a_k^\dagger, \mathcal{H}_k] \quad\text{and}\quad i\hbar\frac{\partial a_k}{\partial t} = [a_k, \mathcal{H}_k].$$

Using identity relations $a_k a_k = 0$ and $a_k^\dagger a_k^\dagger = 0$ in the above, we derive

$$i\hbar\dot{a}_k = \varepsilon_k - a_{-k}^\dagger\left(V\sum_k a_{-k}a_k\right) \quad\text{and}\quad i\hbar\dot{a}_{-k}^\dagger = -\varepsilon_{-k} - a_k\left(V\sum_k a_k^\dagger a_{-k}^\dagger\right). \tag{12.11b}$$

Here, in the original BCS theory, although assumed to be constant, the potential V depends on the modulated lattice, playing a significant role to form a Cooper pair in the soliton theory discussed in chapter 11. Nevertheless, for the ground state, V can be considered as constant real, therefore the following theory is acceptable, while supported by the soliton theory. We can thus define quantities

$$\Delta_k = V\sum_k a_{-k}a_k \quad \text{and its complex conjugate} \quad \Delta_k{}^* = V\sum a_k^\dagger a_{-k}^\dagger.$$

For two eigenvalues $\varepsilon_{\pm k}$ separated from ε_0, corresponding to Fröhlich's condensates for $|\varepsilon_{\pm k}| < \varepsilon_0$. If otherwise $|\varepsilon_{\pm k}| > \varepsilon_0$, we need to take $\Delta_k = \Delta_k{}^* = 0$ for no pair to be formed, hence using the complex Δ_k in (12.11*b*), we obtain

$$i\hbar\dot{a}_k = \varepsilon_k a_k - \Delta_k a_{-k}^\dagger \quad \text{and} \quad i\hbar\dot{a}_{-k} = -\varepsilon_k a_{-k}^\dagger - \Delta_k{}^* a_k. \tag{12.11c}$$

Since operators a_k and a_{-k} have time-factors $e^{\mp i\varepsilon_k t/\hbar}$, respectively, we can solve the following determinantal equation for these eigenvalues ε_{+k} and ε_{-k}, i.e.

$$\begin{vmatrix} \varepsilon_{+k} - \varepsilon_0 & \Delta_k \\ \Delta_k{}^* & \varepsilon_{-k} + \varepsilon_0 \end{vmatrix} = 0 \quad \text{and} \quad \varepsilon_{\pm k}^2 = \varepsilon_0^2 + \Delta_k{}^*\Delta_k. \tag{12.11d}$$

Noticing the above is a familiar level-crossing problem that was discussed in chapter 4, we consider the binary order variable for the Cooper pair $a_{k'}^\dagger a_{k'} = \sigma_{k'}$ at $\boldsymbol{k} = \boldsymbol{k}'$ behaves like a *classical vector*, varying between $+\boldsymbol{k} > \boldsymbol{k}' > -\boldsymbol{k}$, characterized by its finite amplitude. We therefore consider that the corresponding eigenvalue $\varepsilon_{k'}$ represents the pair particle by the operator α_k for the superconducting transition.

Known as *Bogoljubov transformations* [5], we define linear combinations

$$\begin{aligned} \alpha_k &= u_k a_k - v_k a_{-k}^\dagger, & \alpha_{-k} &= u_k a_{-k} + v_k a_k^\dagger, \\ \alpha_k^\dagger &= u_k a_k^\dagger - v_k a_{-k} \quad \text{and} & \alpha_{-k}^\dagger &= u_k a_{-k}^\dagger + v_k a_k, \end{aligned} \tag{12.12a}$$

whose reverse transformations are written as

$$\begin{aligned} a &= u_k \alpha_k + v_k \alpha_{-k}^\dagger, & a_{-k} &= u_k \alpha_{-k} - v_k \alpha_k^\dagger, \\ a_k^\dagger &= u_k \alpha_k^\dagger + v_k \alpha_{-k} \quad \text{and} & a_{-k}^\dagger &= u_k \alpha_{-k}^\dagger - v_k \alpha_k. \end{aligned} \tag{12.12b}$$

In (12.12*a*) and (12.12*b*), u_k and v_k are real coefficients, making symmetric and antisymmetric combinations of one-particle operators with respect to inversion $\boldsymbol{k} \to -\boldsymbol{k}$, i.e. $u_k = u_{-k}$ and $v_k = -v_{-k}$; and normalized as $u_k^2 + v_k^2 = 1$.

For these Bogoljubov's operators, we have significant relations, namely,

$$[\alpha_k, \alpha_{k'}]_+ = u_k u_{k'}[a_k, a_{k'}^\dagger] + v_k v_{k'}[a_{-k}^\dagger a_{-k'}] = \delta_{kk'}(u_k^2 + v_k^2)$$

and

$$[\alpha_k, \alpha_{-k}]_+ = u_k v_k[a_k, a_k^\dagger] - v_k u_k[a_{-k}^\dagger, a_{-k}] = u_k v_k - v_k u_k = 0. \tag{12.12c}$$

Differentiating (12.11*c*) for steady states, we obtain $\varepsilon_{k'}u_{k'} = \varepsilon_k u_k + \Delta_k v_k$. Combining this with (12.11*d*), we have the relation $\Delta_k(u_k^2 - v_k^2) = 2\varepsilon_k u_k v_k$. It is noted that these results suggest a relation to define an angle θ_k to satisfy $\tan\theta_k = \frac{\Delta_k}{\varepsilon_k}$, thereby writing

$$u_k = \cos\frac{\theta_k}{2} \quad \text{and} \quad v_k = \sin\frac{\theta_k}{2}. \tag{12.12d}$$

In addition, we obtain the following relations, i.e.

$$u_k^2 = \cos^2\frac{\theta_k}{2} = \frac{1}{2}\left(1 + \frac{\varepsilon_k}{\varepsilon_{k'}}\right) \quad \text{and} \quad v_k^2 = \sin^2\frac{\theta_k}{2} = \frac{1}{2}\left(1 - \frac{\varepsilon_k}{\varepsilon_{k'}}\right). \qquad (12.12e)$$

For the ground state wave function $|\Phi_o(\boldsymbol{k})\rangle$ of many correlated Cooper pairs, we derive from (12.12c) the expression $\alpha_{-k}\alpha_k|\Phi_{\text{vac}}\rangle = -v_k(u_k + v_k a_k^\dagger a_{-k}^\dagger)|\Phi_{\text{vac}}\rangle$, from which we obtain the correlated density, expressed as

$$\langle\Phi_{\text{vac}}|(u_k{}^* + v_k{}^* a_k a_{-k})(u_k + v_k a_{-k}^\dagger a_k^\dagger)|\Phi_{\text{vac}}\rangle = (u_k^2 + v_k^2)\langle\Phi_{\text{vac}}|\Phi_{\text{vac}}\rangle.$$

Therefore, $|\Phi_o(\boldsymbol{k}')\rangle$ of *correlated multi-particles* is determined by

$$|\,\Phi_o(\boldsymbol{k}')\rangle = (u_{k'} + v_{k'} a_{k'}^\dagger a_{-k'}^\dagger)|\Phi_{\text{vac}}\rangle. \qquad (12.12f)$$

12.4.2 Superconducting state at finite temperatures

Thermodynamically, the BCS theory can be applied at a finite temperature. For finite temperature T, as discussed by Ginzburg–Landau's theory, the ground state $|\Phi_o\rangle$ of Cooper pairs should be excited to temperature-dependent levels by creation operators in such way as

$$\Phi_o\rangle, \quad \alpha_k^\dagger|\Phi_o\rangle, \quad \alpha_k^\dagger\alpha_k^\dagger|\Phi_o\rangle, \ldots.$$

These states should be thermally evaluated by probabilities

$$f_k = 1, \langle\alpha_k^\dagger\rangle, \langle\alpha_k^\dagger\alpha_k^\dagger\rangle, \ldots, \qquad (12.12g)$$

allowing to consider the entropy of the system determined by

$$S = \mathrm{k_B}\sum_k\{f_k \ln f_k + (1 - f_k)\ln(1 - f_k)\}. \qquad (12.12h)$$

We re-express the Hamiltonian (12.12a) for the Gibbs' function in thermodynamic argument, where the kinetic energy is written as

$$\sum_k\langle\varepsilon_k(u_k\alpha_k^\dagger + v_k\alpha_{-k})(u_k\alpha_k + v_k\alpha_{-k}^\dagger)\rangle = \sum_k\varepsilon_k\left\{v_k^2(1 - f_k) + u_k^2 f_k\right\}, \qquad (12.12i)$$

noticing that $\alpha_k^\dagger\alpha_{-k}^\dagger = 0$. Here, the first term on the right expression indicates there can be no particles in v_k, depending of the probability f_k, while all others are in u_k; f_k and $1 - f_k$ represent the statistical weights on average numbers in these modes.

The interacting potential can be calculated as

$$\sum_{k'} V(\boldsymbol{k}, \boldsymbol{k}'; \boldsymbol{q}, -\boldsymbol{q})(1 - 2f_k)(1 - 2f_{k'})$$

which is factorized as follows, if minimized with respect to $\boldsymbol{k}'$, i.e.

$$V(\boldsymbol{k}, \boldsymbol{k}'; \boldsymbol{q}, -\boldsymbol{q}) \rightarrow V_k(\boldsymbol{q})V_{-k}(-\boldsymbol{q}). \qquad (12.12j)$$

Gibbs' function can then be expressed as (12.12*k*). Therefore, as (12.12*e*) was derived, we expressed that

$$u_k^2 = \frac{1}{2}\left(1 + \frac{\varepsilon_k}{\sqrt{\varepsilon_k^2 + \Delta_k(T)_2}}\right) \quad \text{and} \quad v_k^2 = \frac{1}{2}\left(1 - \frac{\varepsilon_k}{\sqrt{\varepsilon_k^2 + \Delta_k(T)^2}}\right). \qquad (12.12k)$$

Minimizing Gibbs' function (12.12*k*) with respect to f_k, we obtain the relation

$$\mathrm{k_B}T\{\ln f_k + \ln(1 - f_k)\} + \sqrt{\varepsilon_k^2 + \Delta_k(T)^2} = 0,$$

leading to the formula

$$f_k = \frac{1}{1 + e^{E_k/\mathrm{k_B}T}} \quad \text{where} \quad E_k = \sqrt{\varepsilon_k^2 + \Delta_k(T)^2} \qquad (12.12l)$$

Here, the function f_k represents the *Fermi–Dirac distribution* of fermion particles.

Exercises

1. Meissner's perfect diamagnetism is proposed based on the externally applied magnetic field ***H*** penetrated through the container of experimental device, unlike the applied pressure. Discuss this experimental issue to determine if this property of ***H*** is either empirical or a law of nature?
2. To observe Ginzburg–Landau's dynamics, we need a Weiss' potential. How we can find it? Discuss the physical logic behind this problem.
3. Equation (12.12*l*) is a proof for Cooper pairs and Fröhlich's solitons to obey Fermi- and Bose-statistics, respectively. Regarding the derivation of Weiss' potential, compare (12.12*l*) with the discussion in chapter 11, to satisfy the mechanism of superconducting transition.

References

[1] Kuper C G 1968 *An Introduction to the Theory of Superconductivity* (Oxford: Clarendon)
[2] Kittel C 1986 *Introduction to Solid State Physics* (New York: Wiley)
[3] Ginzburg V L and Landau L D 1950 *Zh. Eksp. Teor. Fiz.* **20** 1064
[4] Bogoljubov N N 1947 *J. Phys. Moscow* **11** 23
[5] Bogoljubov N N 1958 *Nuovo Cimento* **X7** 794

CPSIA information can be obtained
at www.ICGtesting.com
Printed in the USA
BVHW011042110122
625571BV00025B/18

9 780750 337571